Italian Aviation Units in the First World War

Volume 1

Roberto Gentilli

ISBN: 978-1-953201-90-4

Text edited by: Colin Owers
Cover Painting: Thomasz Sadzinica
Color Profiles: Bob Pearson
Design and layout: Jack Herris
Cover design: Aaron Weaver
Digital photo editing: Jack Herris

Aeronaut Books
Books for Enthusiasts by Enthusiasts
www.aeronautbooks.com

Table of Contents

Volume 1

Cover Painting: Two Caproni bombers on a daylight bombing raid are intercepted by Austro-Hungarian Albatros D.III fighters.

General Chronology

1914

28 July - Austria declares war on Serbia, the Great War begins.
3 August – Italy announces its neutrality.

1915

26 April – The secret Pact of London is signed, ensuring the entry of Italy into the war.
3 May – Italy denounces its Treble Alliance with Austria and Germany and begins mobilization.
24 May – Italy declares war on Austria.
23 June – 7 July – First Battle of the Isonzo.
18 July – 3 August - Second Battle of the Isonzo.
18 October – 4 November – Third Battle of the Isonzo.
10 November – 2 December – Fourth Battle of the Isonzo.

1916

9 – 15 March – Fifth Battle of the Isonzo.
14 May – The Austrians launch a large offensive from the Asiago plateau.
25 June – An Italian counter-offensive regains some lost ground in the North.
6 – 17 August – Sixth Battle of the Isonzo. Conquest of Gorizia.
27 August – Italy declares war on Germany.
14 – 17 September – Seventh Battle of the Isonzo.
10 – 12 October – Eighth Battle of the Isonzo.
31 October – 4 November – Ninth Battle of the Isonzo.

1917

6 April – The United States declares war on Germany.
12 May – 5 June – Tenth Battle of the Isonzo
10 June – Battle of the OrTigara.
17 – 31 August – Eleventh Battle of the Isonzo.
24 October – Twelfth Battle of the Isonzo – Austro-German breakthrough at Caporetto.
8 November – The king of Italy meets with French prime minister Painlevé and Britsh premier Lloyd George and promises Italy's permanence in the war. Armando Diaz replaces Luigi Cadorna as supreme commander.
7 December – The United States declares war on Austria.

1918

15 June – The Austrians launch a general offensive across the River Piave, that fails.
24 October – The final offensive, the Battle of Vittorio Veneto, begins.
3 November – Austria accepts an armistice, with hostilities ending the enxt day.

Part 1: Preface

This work is not conceived as a general history of Italian aviation in the Great War, but as a collection of data, complete and exact as much as possible, on the single units, Gruppi, Squadriglie and Sezioni, a database of names, serial numbers, dates and events.

As a "database" therefore this book is designed as a source of consultation. Offering many tens of thousands of data provided through a careful study of archival sources, we think it comes close to the goal of supplying an objective and complete base for future research on Italian aviation in the First World War. This book, having Paolo Varriale as co-author, was originally published in Italian in 1999 by the *Ufficio Storico dell'Aeronautica Militare*, the Historical Office of the Italian Air Force, as "*I Reparti dell'aviazione italiana nella Grande Guerra*". In the following quarter century, it has been amended and integrated, now including also the first two years of naval aviation activities and the flying schools.

This work, however, is not new, because it only follows the path of the detailed research on combat squadrons of the Army done in the inter-war years by Colonel Ercole Morelli of the Historical Office of the Regia Aeronautica. What differentiates this from Morelli's research, besides the inclusion of minor units and of the Navy squadrons, is the availability of many Austrian, German and British sources and the power of Internet, but in every stage of this work I was amazed by the completeness and accuracy of the papers of Morelli, to whom I posthumously express my gratitude, as much as to the Ufficio Storico dell'Aeronautica Militare that continues a precious program of classification and analysis.

Although the goal was a work of reference, I tried to add a human aspect in these pages, recalling events that now are forgotten, and to share with the reader the fascination of a heroic period and of so many men serving their country. It was the human aspect that made this work rewarding and agreeable: not just the presentation of the famous airmen, aces like Baracca and Scaroni, and commanders like Bonazzi, Armani and Pierozzi, but above all the unexpected discovery of men who gave so much, flying hundreds of difficult and unsung missions, and who were ignored at the time, like *tenente* Bortolotti, *maresciallo* Pancani, *sergenti* Stillio and Sirigatti, and so many more, airmen whose only reward was the appreciation of their partners. The index, that lists over 4,400 airmen with the units in which they served and their decorations, is the main tool that this book offers, and its main *raison d'être*.

Readers should be warned, however, that this research, although painstaking, is based on documents that are far from complete, and that in many cases were somewhat incorrect to start with. Airmen at the front had many tasks, and record keeping was one of them, but certainly not their priority. All information, including names, dates, serial numbers, ranks, has been checked as much as possible, but readers should be aware that this work cannot claim to be exhaustive or faultless. Especially in trying to match Italian losses with Austrian air claims, one should consider that much data, particularly of the units of the crucial II and VI Gruppi before October 1917, are missing, and there are elements pointing to more combat losses than it has been

Rank Equivalance

Italian Army	Italian Navy	English (Army)	English (Navy)	German
Generale	Ammiraglio	General	Admiral	General
Colonnello	Capitano di vascello	Colonel	Captain	Oberst
Tenente colonnello	Capitano di fregata	Lieutenant colonel	Commander	Oberstleutnant
Maggiore	Capitano di corvetta	Major	Lt. Commander	Major
Capitano	Tenente di vascello	Captain	Lieutenant	Hauptmann
Tenente	Sottotenente di vas.	First lieutenant	Jr. Lieutenant	Oberleutnant
Sottotenente	Guardiamarina	Second lieutenant	Ensign	Leutnant
Aspirante			Midshipman cadet	Fähnrich
Aiutante di battaglia		Chief w. officer	Chief w. officer	
Maresciallo		Warrant officer	Warrant officer	Oberfeldwebel
Sergente maggiore	2° capo	Sergeant major	Chief petty officer	Feldwebel
Sergente	Capo	Sergeant	Petty officer	Unteroffizier
Caporale	Sottocapo	Corporal	Seaman 1 Cl.	Korporal
Soldato	Marinaio	Private	Seaman	Flieger

possible to identify and report in these pages.

My warmest thanks are due to the Ufficio Storico dell'Aeronautica Militare and to the Ufficio Storico della Marina Militare. I also thank for their help first of all our friends Luigino Caliaro, Fiorenzo Longhi and Karl Meindl, unequalled in their knowledge of aviation and in generosity, and Gregory Alegi, Mauro Antonellini, Giorgio Apostolo, Angelo Emiliani, Antonio Iozzi, Alessio Meuti, Colin A. Owers and then: Andrei Alexandrov, Aldo Astolfi, Gianandrea Bussi, Renato Cerutti, Mario Chirico, Boris Ciglic, Zoltàn Cziròk, Massimiliano De Antoni, Albin Denis, Tadeusz P. Dobrowiecki, Eugenio Fabris, Paolo Farina, Mario Federighi, Ermanno Ghidoi, Cesare Gori, Rudolf Hoefling, Paolo Miana, Pier Luigi Moncalvo, Angelo Moneta, Paolo Monti, Lucien Morareau, Gaetano Pastoressa, Gino Piccoli, William Rabito, Jiri Railich, Ray Rimell, Pasqualino Schifano, Walter Schroeder, Luigino Scroccaro, Umberto Scricciolo, Mirko Sernaglia, Dina Tamiozzo, Achille Vigna, Richard Ward, Jan Zahalka, some of them no longer with us.

Abbreviations: pil. pilota, oss. osservatore, mitr. mitragliere (gunner), sold. soldato, cpr. caporale, cpr.m. caporal maggiore, serg. sergente, serg.m. sergente maggiore, m.llo maresciallo, a.b. aiutante di battaglia, carab. carabiniere, brig. brigadiere, asp. aspirante, s.ten. sottotenente, ten. tenente, cap. capitano, magg. maggiore, t.col. tenente colonnello, marò marinaio, torp. torpediniere, cann. cannoniere, nocc. nocchiere, s.capo sottocapo, GM Guardiamarina, STV Sottotenente di Vascello, TV Tenente di Vascello, CC Comandante di Corvetta, CF Comandante di Fregata, CV Comandante di Vascello, RR.CC. Reali Carabinieri.

Localities are indicated with the Italian names used during the war, the current names of localities now part of Croatia, Slovenia, Montenegro and Albania are different.

Above: A large parade was organized on 20 September 1918 at Padua airfield. Behind a first row of three Spads and six SVAs, there are Capronis with different one or two colors camouflage over the wings.

The Italian front : from 1915 until 24 october 1917

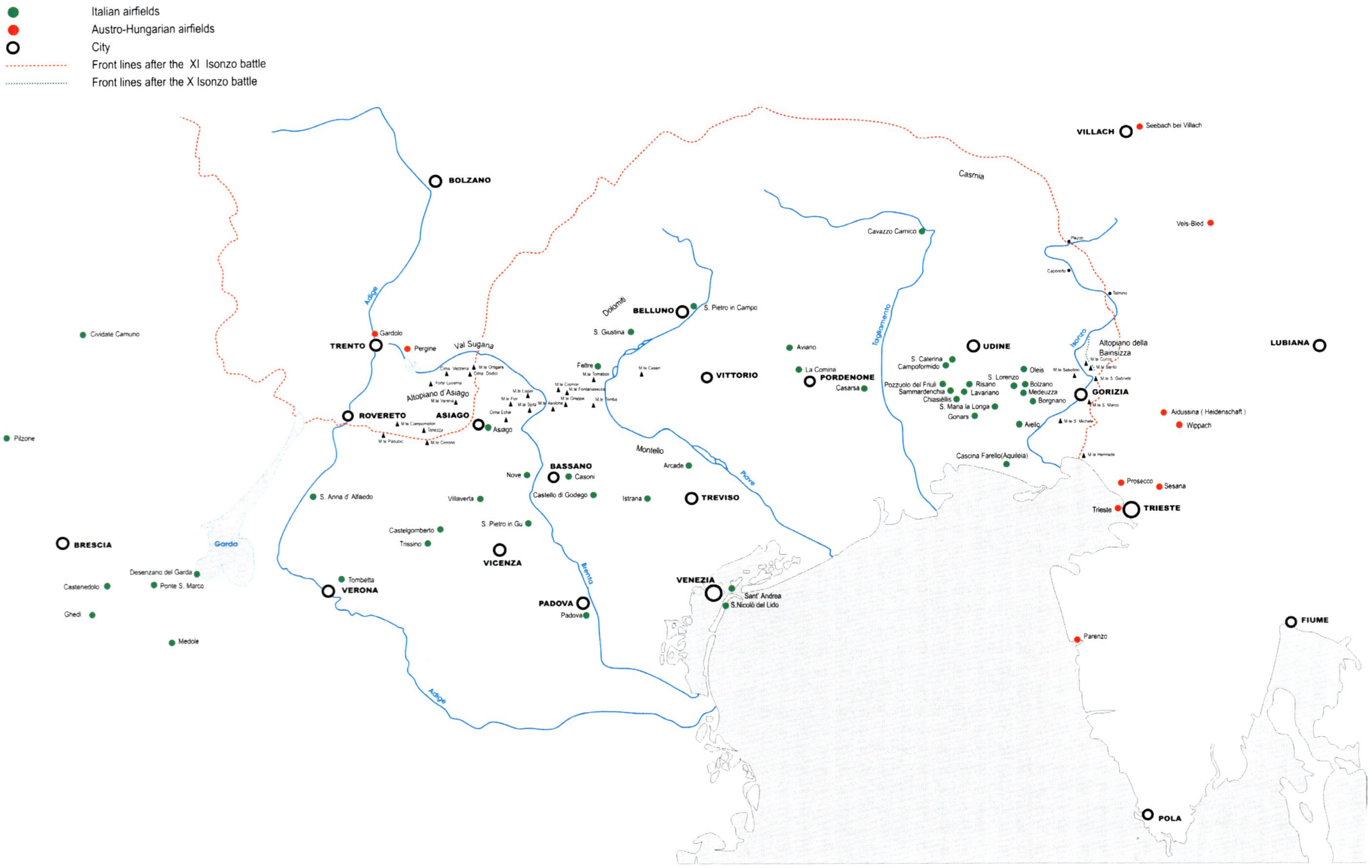

The Italian front : November 1917 October 1918

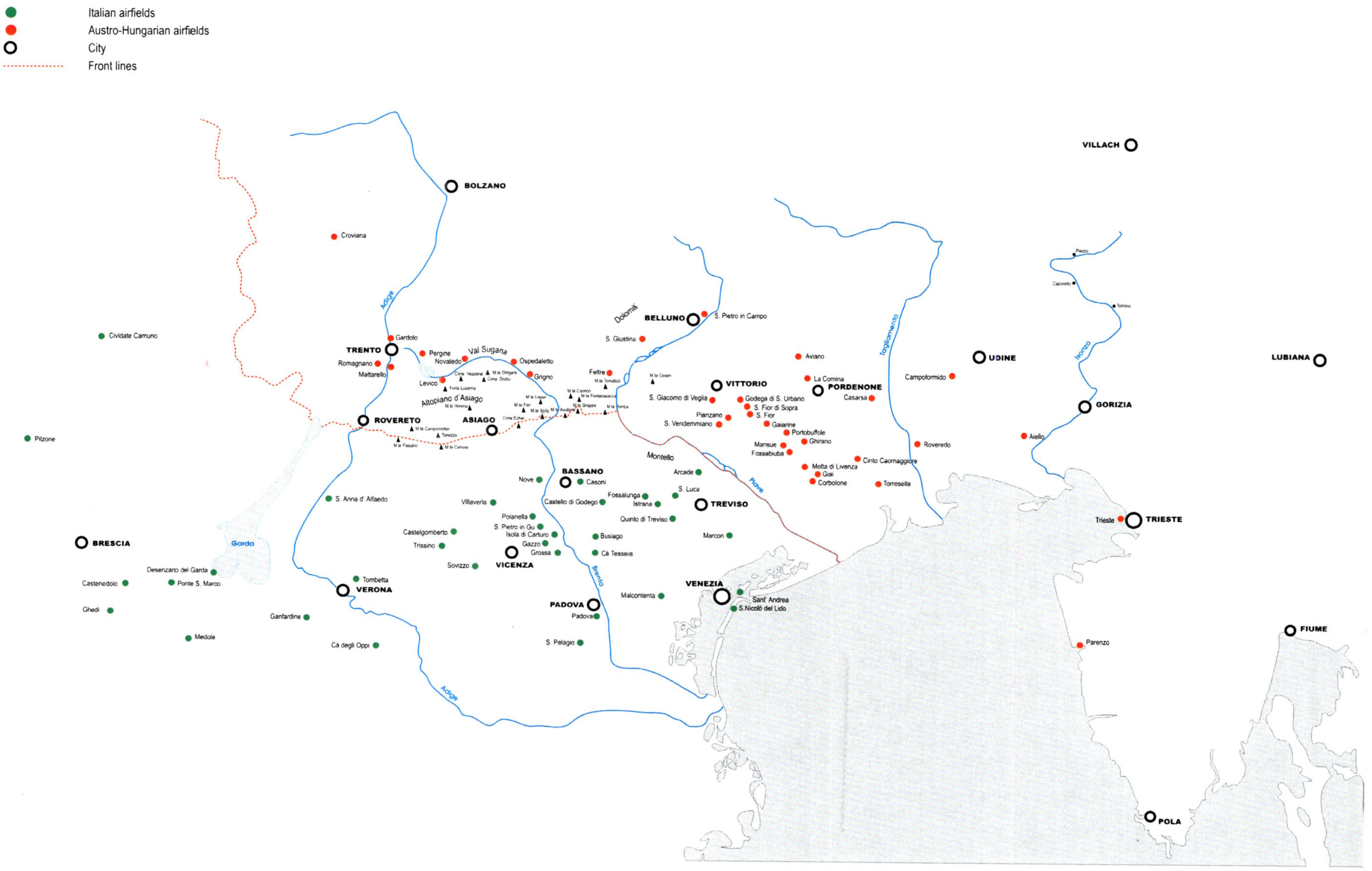

Part 2: Introduction

The Italian Air Force in the First World War

Above: Uniforms of Italian aviators at the start of the Great War in a painting by the reputed illustrator Quinto Cenni.

The air force was born in Italy in 1911, as a part of the Battalion of Engineering Specialists, and already from October of that year it had its first military use in Libya. A law in June 1912 created, under the General Directorate of Artillery and Engineering, an Aeronautical Service formed by the Aviators Battalion, with command at the La Marmora Barracks in Turin and base in Mirafiori, Turin's airfield, under the command of *ten. col.* Vittorio Cordero di Montezemolo, who was entrusted with the management of the airplanes, while the Specialists Battalion in Rome remained in charge of lighter than air: airships and balloons.

In 1913 commander of the Aviators Battalion became Major Giulio Douhet, future theorist on the command of the air. At the end of 1913 there were 109 military pilots, all officers except for twelve non-commissioned officers, twelve squadrons with about 80 Blériot, Nieuport, Farman and Bristol aircraft and four military pilot schools, in Pordenone, Mirafiori, Malpensa and Aviano.

Like France in 1911, Germany, Great Britain and Russia in 1912, Italy too in 1912 announced a Military Aviation Competition, managed by a commission chaired by *magg.* Douhet, with a tender for an airplane built in Italy, even if based on a foreign design, another, for an airplane built and designed in Italy, which included the order for seven aircraft as a prize. The requirements were: two seats, a range of at least 300 km and provision for 40 kg of bombs on board. The trials took place in April 1913 at Mirafiori, and sixteen aircraft were admitted, all monoplanes except two SAML biplanes. The contest, which also saw a fatal accident, did not lead to any results: the Bobba monoplane won but was not ordered, leading to a long litigation.

In the last years of peace throughout Europe, aviation, far from being considered an eccentricity for sportsmen, was recognized as a new weapon with vast potential, both by public opinion and by military leaders, indeed there was almost an exaltation, a patriotic paroxysm reinforced by the frequent international meetings and air races.

In 1913 a national subscription was organized in Italy by the National Air League "Pro aerial fleet" to "give wings to Italy" which yielded 3,697,000 lire, equal to 15,000,000 dollars, a considerable sum in every era. This sum was spent in 1915, allowing the purchase at a cost of 2,747,000 of 102 airplanes (Blériot, Nieuport, Farman and Caproni 300 hp), while the remaining 900,000 lire were later used to purchase some Caproni 450 hp. In 1914 various decrees consolidated the structure of the Italian air force, promoted

Above: The Macchi Parasol was the first airplane of Italian design in service, but it was unstable and dangerous, and was soon retired.

to Aeronautical Inspectorate. Army Chief of Staff, General Alberto Pollio, was reluctant to commit funds to a rapidly evolving sector, but the real problem with underdevelopment in 1913–14 was lack of Government funds after the high costs of the Italo-Turkish war.

At the outbreak of the European war, the Italian Air Force did not lack good leaders and excellent officers, and had a well-structured organization, which ranged from the Stabilimento di Esperienze e Costruzioni Aeronautiche (Aeronautical Construction and Experience Establishment) in Rome, to the operational squadrons and three pilot schools, with good training: it lacked the tools to deal with the exponential increase in tasks that war would bring, especially an industrial base. But the months of neutrality were not wasted, and were used to obtain manufacturing licenses in France and lay the foundations for the development of national industry.

A Royal Decree dated 7 January 1915, established the Corpo Aeronautico Militare (Military Aeronautical Corps), CAM, ordered with two commands, one for aviators and one for the lighter-than-air, a Direzione Generale d'Aeronautica, (General Directorate of Aeronautics), DGA, with a Battaglione Squadriglie Aviatori (Squadrons Battalion), a Battaglione Scuole Aviatori (Schools Battalion) and a Direzione Tecnica dell'Aviazione Militare, (Military Aviation Technical Directorate), DTAM, alongside the Battaglione Dirigibilisti (Airshipmen Battalion) and the Battaglione Aerostieri (Aerostat Battalion). The Royal Decree (it was converted into law only in 1917) also established the civilian technical specialist staff for the Air Force, made up of engineers, professors and specialists, while a Central Aeronautical Institute replaced the Stabilimento di Esperienze e Costruzioni Aeronautiche with technical-scientific tasks, and preparation of new construction projects. Colonel Maurizio Moris was at the head of the DGA, while Major Carlo Piazza was chief of Battaglione Squadriglie Aviatori, and at the head of the DTAM was Major Ottavio Ricaldoni.

The budget allocated for the 1914-1915 fiscal year was 4,145,000 lire, and included an extraordinary allocation of 16,500,000 lire (65,000,000 dollars today), of which 5,000,000 for the Ministry of the Navy.

More had been spent in other countries, but the outbreak of war in August 1914 found all the various air forces in a period of transition between the monoplanes of the pioneer era (such as the Blériots, REPs and Morane Saulniers in France, and the Taube in Germany) and the biplanes, a transition that was similar to the one encountered by Italy ten months later. The uncertainties about the organization were common: Germany, for example, had equipped itself in 1913 with five aviator battalions which depended, through an Inspectorate, on the Inspector General of Military Transport.

What form warplanes would take was still an uncertain guess. The Germans opted for biplanes with in-line frontal engines, and the choice proved to be right, but for the first period of the war, operating the Taube monoplanes, their inferiority compared to the French was marked. The latter focused on the wrong solution, the rear-engined tailbeam biplane, whose capabilities as a war machine were already exhausted by the end of 1916. The aircraft that showed the way for almost all manufacturers was the English Sopwith Tabloid, a small biplane with clean lines, which in seaplane version had won the Schneider trophy in 1914.

On the eve of the war, the Aviator Squadrons Battalion was made up of three Gruppi (Groups), with authority and functions similar to those of a battalion, and fifteen Squadrons, corresponding to companies. The airplanes were about eighty, not one more than in 1913: 37 Blériots, 27 Nieuports, 22 Farmans. This weakness of the Italian aviation was later criticized, but considering the two factors, the scarce budgets

Italian Air Units on 24 May 1915

Comando Battaglione Squadriglie Aviatori – Pordenone		magg. Alfredo Barbieri
Reporting to 3a Armata		
I Gruppo,	Portogruaro	cap. Novellis di Coarazze
1ª Blériot	Portogruaro	cap. Costantino Quaglia
2ª Blériot	Portogruaro	cap. Ercole Capuzzo
3ª Blériot	Portogruaro	cap. Egidio Carta
13ª Blériot	Torresella	cap. Vincenzo Lombard
14ª Blériot	Torresella	cap. Tullio Visconti
Reporting to 2a Armata		
II Gruppo,	Campoformido	cap. Lelio Gaviglio
6ª Nieuport	Campoformido	cap. Oronzo Andriani
7ª Nieuport	Campoformido	cap. Armando Armani
8ª Nieuport	Campoformido	cap. Adolfo Resio
Reporting to Comando Supremo		
III Gruppo	Pordenone	cap. Augusto Gallina
5ª Nieuport	Campoformido	cap. Pier Ruggero Piccio
9ª Farman	Pordenone	cap. Giorgio Chiaperotti
10ª Farman	Pordenone	cap. Carlo Tappi
Reporting to Comando Piazza Marittima Venezia		
4ª Blériot	Bazzera	cap. Guido Olivo
In trasformazione		
11ª Farman	Brescia	cap. Michele Masi
12ª Farman	Verona	cap. Ferdinando De Masellis
15ª Caproni 80 cv	Piacenza	cap. Umberto Rossi.

assigned in 1913 and 1914, when the extraordinary effort of the Libyan war still weighed (nor was there any reason to think that an archduke would have been killed in Sarajevo) and the weakness of Italian industry, still dependent on French patents and on imports even from Germany, England and the United States, it is more correct to recognize that with the little available they had worked well.

Order of Battle August 1915

Sezione Caproni	Pordenone
2a Armata	
II Gruppo	Udine
2a Sq. Farman	Campoformido
9a Sq. Farman	Campoformido
2a Sq. Artiglieria Macchi	Medeuzza
3a Armata	
I Gruppo	Cervignano
5a Sq. Voisin	Campoformido
11a Sq. Farman	Chiasiellis
8a Sq. Nieuport (II Gr.)	Clauriano
III Gruppo	S. Maria la Longa
6a Sq. Nieuport	S. Maria la Longa
10a Sq. Farman	S. Maria la Longa
1a Sq. Artiglieria Caudron	Medeuzza
1a Armata	
12a Sq. Farman	Asiago
Sezione Idrovolanti	Desenzano
Comando di Venezia Regia Marina	
4a Blériot	Bazzera

Although Italy was allied with Germany and Austria, in the so-called Triple Alliance dating from Franco-Italian rivalry over Tunisia in 1882 and renewed in 1912, the country had no reason to oppose Britain and France in the Twentieth century, while the Austro-Hungarian Double Monarchy still ruled the two Italian-speaking provinces of Trento and Trieste. Thus, Italy remained neutral in 1914 when the European war began, as article 4 of the Alliance allowed.

In December 1914, when it was clear that Italy would not side with the former allies of the Triple Alliance, some squadrons were moved East, from Piedmont and Lombardy to Aviano and Pordenone, while the schools there were moved to Mirafiori and Somma Lombardo (Malpensa). Special courses for observer officers were also hastily created. The recruitment of volunteers in possession of a pilot's license was also regulated, which brought excellent pilot soldiers into the ranks of the Corps, who then needed a long process to become second lieutenants. The Navy did the same with the owners of motor boats, called Volontari Motonauti.

Public opinion was divided: Catholics, Socialists and Moderates were for neutrality (although for different reasons) while Republicans, Radicals and Nationalists (also for quite diverging reasons) were for intervention on the Allied side. Leftist Socialist leader Benito Mussolini had a conversion to the far right and became a proponent of war. Great Britain and France promised help and territorial gains to Italy if it would join the Allies in war and on 26 April

Above: Caproni 300 hp bombers taking off from Aviano in February 1916.

1915 the secret Pact of London was signed, promising Italian supremacy in the Adriatic Sea in exchange for its entry in the war. Poet Gabriele d'Annunzio made speeches and organised demonstrations for intervention, and on 24 May 1915 Italy became a belligerent, declaring war on Austria (war to Germany was declared only in August 1916).

Italy Enters the War

The Army, under the supreme command of General Luigi Cadorna, deployed at the front, that stretched from the Tyrolean Alps to the Adriatc Sea, with its 3ª Armata (Army Corps VII, X e XIV) south, from the sea up to Pozzuolo del Friuli, the 2ª Armata (Army Corps II, IV and XII) from Udine up to Gemona and Tarcento, the Brigades *Piemonte* and *Aosta* in the Carnia region, the 4ª Armata in the northern sector from Cadore to Belluno, and the 1ª Armata covering the northern front of Veneto, from Feltre to the Asiago plateau.

At mobilisation the Command of the Battaglione Squadriglie Aviatori started operating with the Supreme Command, establishing a strict connection between aviation and Army that proved fruitful during the war. Its commander was *col.* Giuseppe Motta.

The aviation force of the Italian Army consisted of six Bleriot squadrons with 37 aircraft and 30 pilots, four squadrons of Nieuport monoplanes with 27 aircraft and 20 pilots, four Farman squadrons with 22 aircraft and 20 pilots.

In June the Comando Squadriglie Aviatori moved to Pordenone. The III Gruppo was assigned to the Supreme Command, the II Gruppo to the 2ª Armata, the I Gruppo to the 3ª Armata, the 1ª Armata, in the northern part of the front, and the 4ª Armata, in Cadore, received no aviation unit.

The I Gruppo went to Mortegliano with the 1ª, 2ª and 3ª Bleriot at Chiasiellis, the 13ª and 14ª at Pozzuolo del Friuli. The III Gruppo took in charge also the 6ª Nieuport and moved all its units, 5ª, 6ª, 9ª and 10ª to Campoformido.

In the first days of war Italian troops advanced and conquered Gradisca and Monfalcone, placing a bridgehead across the Isonzo River at Plava, while the Alpini (mountain troops) of the 2ª Armata stormed the Monte Nero but the opportunity to complete a much deeper advance against still minimal Austrian defences was missed. From 23 June to 7 July the 1st Battle of the Isonzo was fought, the Italians attacked frontally the Austrian positions on the Carso, a rocky unhospitable plateau, without gaining any ground. The national markings of black circles and bars on the wings was replaced with the colours of green and red, to be applied to the wings of airplanes at midnight, July 15.

On 18 July the 2nd Battle of the Isonzo started, it ended on 4 August again without any shift of the lines. During the summer the Squadriglie with the old Bleriot and Nieuport monoplanes were disbanded. First to be disbanded was the 1ª and 14ª in June, then in July the 2ª Bleriot and 5ª Nieuport, then the 8ª Nieuport, while new units with more effective airplanes were in formation. In order to meet the desire of the Arm of Artillery to have its own units, the 1ª and 2ª Squadriglia Artiglieria were formed, equipped respectively with the Caudron G.3 and the Macchi Parasol, forming a Gruppo Squadriglie Aviatori per l'Artiglieria. In mid July came the new 1ª Squadriglia with Farman type 1914. Two new squadrons were receiving the Voisin LA, a French design built under licence, slow but strong and reliable; they were to be numbered 16ª and 17ª, but they received the numbers 5ª and 7ª, as these were now available. In late July a Sezione (flight) armed with Caproni 300 trimotor bombers went into formation at Pordenone. On 7 August the 1ª Farman squadron arrived at the front, with the new Maurice Farman 1914, first with 80 hp De Dion engines, then 100 hp Fiats. Although an archaic and slow biplane it had excellent flight qualities and would remain in service until 1919.

On 27 August 1915 the confusion of unit numbers was removed, unifying the numbers and assigning to the new Farman, Voisin and Aviatik squadrons the numbers of the disbanded Bleriot, Nieuport and Farman 1912 units. The Order of Battle in September reflected the situation of the Italian air force, at its lowest level ever, due to the retirement of the obsolete models: it consisted of just I Gruppo 11ª Farman, 5ª and 7ª Voisin, II Gruppo 9ª Farman, III Gruppo 10ª Farman.

The III Gruppo was disbanded, giving its single squadron to the I Gruppo. On 18 October 1915 a new offensive began,

Typical apparel of an Italian pilot.

Order of Battle 31 December 1915

Comando Battaglione Sq. Aviatori, Pordenone		ten. col.Barbieri
1ª Sq. Offesa Caproni	Pordenone e Aviano	cap. Bailo
2ª Sq. Offesa Caproni	" "	cap. Carta
3ª Sq. Offesa Caproni	" "	cap. Piccio
4ª Sq. Offesa Caproni	" "	cap. Ungania
6ª Sq. Offesa Caproni	" "	cap. Resio
Gruppo Squadriglie per difesa antiaerea di Udine		
2ª Sq. Farman	Campoformido	cap. Masi
1ª Sq. Caccia Nieuport	S. Caterina	cap. Tacchini
Comando I Gruppo, S.Maria la Longa, 3ª Armata		magg. Gallina
5ª Sq. Voisin	Chiasiellis	cap. De Rada
7ª Sq. Voisin	S. Maria la Longa	cap. Capuzzo
10ª Sq. Farman	S.Maria la Longa	cap. Tappi
11ª Sq. Farman	Chiasiellis	cap. Varanini
Comando II Gruppo, Udine, 2ª Armata		magg. Gaviglio
6ª Sq. Farman	Chiasiellis	cap. Olivo
9ª Sq. Farman	Campoformido	cap. Beltramo
Comando III Gruppo, Verona, 1ª Armata		cap. La Polla
5ª Sq. Offesa Caproni	Verona	cap. Andriani
1ª Sq. Farman	Verona	cap. Bonazzi
12ª Sq. Farman	Asiago	cap. Armani
3ª Sq. Caccia Aviatik	Brescia	cap. Matteucci
4ª Sq. Caccia Aviatik	Verona	cap. Coppini
Sezione Idrovolanti	Desenzano	ten. Cesaroni
Comando Gruppo Aviazione per Artiglieria		ten. col. de Siebert
1ª Sq. Caudron	Oleis - 1ª Armata	mag. Gamerra
2ª Sq. Caudron	Risano - 3ª Armata	cap. Santi
3ª Sq. Caudron	Gonars - 3ª Armata	cap. Maurel
4ª Sq. Caudron	Gonars - 3ª Armata	cap. Costanzo
5ª Sq. Caudron	Oleis - 2ª Armata	cap. Lampugnani

the 3rd Battle of the Isonzo that lasted until 4 November, immediately followed by the 4th, until early December, two useless and bloody attempts at breaking through the Austrian defence lines. Counteracting the Austrian success against the Serbian Army, in December Italian troops landed in Albania, opening a new and difficult front, where the Austrians occupied the port of Durazzo.

On 1 November 1915 the Comando Battaglione Squadriglie Aviatori with *t.col.* Barbieri took the role also of *Comando di Gruppo Squadriglie del Comando Supremo*, group command for the squadrons assigned to the Supreme Command, which were the new 1ª, 2ª, 3ª and 4ª Caproni and the Sezione Aviatik da Difesa formed at Aviano. On 1 December 1915 the designations of the Squadriglie were changed, adding the role of each, so they became: Offesa (offence), Ricognizione e Combattimento (reconnaissance and combat), Artiglieria (artillery) and Caccia (fighter). Thus, the 8ª Nieuport became 1ª Squadriglia Caccia. The Sezione Aviatik of Aviano became 4ª Squadriglia Caccia, the 3ª Aviatik was now 3ª Caccia. For Ricognizione e Combattimento numbers 3, 4 and 8 remained vacant. The 2ª Farman and the 1ª Caccia formed a *Gruppo Squadriglie per la Difesa di Udine*, (Udine defence group) with *cap.* Masi of the 2ª as commander. On 1 December the III Gruppo was reformed with Squadriglie 1ª and 12ª Ricognizione, 3ª and 4ª Caccia, 5ª Offesa and the Sezione Idro. This Gruppo supported the Trentino sector of the front, and remained there throughout the war.

Control of the air force was assigned to the *Direzione Generale d'Aeronautica* of the Ministry of War, headed by Col. Maurizio Mario Moris, a supporter of airships. He was replaced in December 1915 by General Giovanni Battista Marieni, a hard-working and able technician, who remained two years and then became Inspector of the Corps of Engineers. The workers of the aeronautical industries increased from 60 to 1,500 in 1915. New factories were rapidly established for the construction of aircraft, engines and accessories. Flying schools were enlarged (for which 900

Above: A Caproni Ca.3 with unusual markings, two question marks, and indeed the squadron to which it belonged is unknown. (Archive Caliaro)

applications arrived) and industrial agreements were made with France, where pilots were sent to learn about the new models. The preparation of good observers was essential, and a special school was therefore set up in Centocelle, Rome, but for the entire duration of the war other courses were organized directly at the squadrons at the front, where the officers could train, be evaluated as students and obtain the appointment as aircraft observer.

As winter came, the front stabilised, and the conflict became stationary trench warfare. In just over seven months of war, the air force had completed about a thousand missions, of which 214 were bombing missions, not big numbers, but which represented gaining experience, in which the important thing was to learn the lessons for a new type of war.

The 5th Battle of the Isonzo began on 11 March 1916, launched in order to engage Austrian forces so as to deny them the possibility to send troops to help the Germans, then fighting the Battle of Verdun. The offensive of the Italian 3ª Armata, stretching from Mount Sabotino to the coast, ended on 30 March without any result.

On 8 April 1916 the numbers issued to Squadriglie were radically changed, creating a unified system that lasted until current times. Numbers from 1 to 24 were reserved for bomber squadrons, those from 25 to 40 for reconnaissance squadrons, from 41 to 69 for artillery squadrons, and from 70 onward to fighters. The *Squadriglie da offesa* 1ª to 7ª kept their number, but *Squadriglie da Ricognizione e combattimento* 1ª, 2ª, 4ª, 5ª, 6ª, 7ª, 10ª, 11ª, 12ª and 13ª became respectively 31ª, 33ª, 29ª, 25ª, 30ª, 26ª, 28ª, 29ª, 32ª and 34ª. The seven *Squadriglie Artiglieria* from 1ª to 7ª became respectively 41ª, 42ª, 43ª, 44ª, 45ª, 46ª and 47ª, and the five fighter Squadriglie took up the numbers 70ª, 71ª, 72ª, 73ª and 74ª.

The I Gruppo was now formed by 25ª, 26ª, 27ª, 28ª and 71ª Squadriglie, the II Gruppo by 29ª and 30ª, the III Gruppo by 31ª, 32ª, 72ª, 73ª, 5ª Caproni and 1ª Idro, the Gruppo Comando had the Squadriglie Caproni 1ª, 2ª, 3ª, 4ª, 6ª and 7ª, while the *Gruppo Aviazione Artiglieria* had the Squadriglie 41ª to 47ª. At the same time progressive numbers were also issued to all Gruppi, then the Gruppo Comando became IV. The Gruppo Aviazione per l'Artiglieria became a Command, and its unit formed the new V Gruppo. Two weeks later, on 1 May 1916, Gruppi VI and VII were also formed, the former with 45ª and 47ª, the latter with 46ª and 48ª.

On 14 May 1916 the Austrians launched their "*Frühjahrsoffensive*" which Italian propaganda called "*Strafexpedition*" (punishing expedition), an offensive coming from an unexpected sector, the Trentino, aiming at the invasion of the Asiago plateau, driving down to Vicenza and to the defenceless plain of the Po, striking at the back of the main Italian front, according to a plan long studied by the Austrian supreme commander *Feldmarschall* Franz Conrad von Hötzendorf. Two Austrian Armies supported by artillery fire that, according to the Austrians, was the heaviest ever unleashed in any war, forced the 1ª Armata to retreat in

Order of Battle 1 April 1916

Units	Base	Aircraft	Qty.
Gruppo Squadriglie per il Comando Supremo			
1ª Sq. Offesa	Pordenone	Caproni 300	2
2ª Sq. Offesa	Pordenone	Caproni 300	3
3ª Sq. Offesa	Pordenone	Caproni 300	3
4ª Sq. Offesa	Pordenone	Caproni 300	2
6ª Sq. Offesa	Pordenone	Caproni 300	4
7ª Sq. Offesa	Aviano	Caproni 300	3
I Gruppo Sq.	**S.Maria la Longa, 3ª Armata**		
5ª Sq. Ric/Comb.	Chiasiellis	Voisin	5
7ª Sq. Ric/Comb.	S.Maria la Longa	Voisin	8
10ª Sq. Ric/Comb.	S.Maria la Longa	Farman	7
11ª Sq. Ric/Comb.	Chiasiellis	Farman	8
2ª Sq. Caccia	Cascina Farello	Nieuport	8
II Gruppo Sq.	**Udine, 2ª Armata**		
4ª Sq. Ric/Comb.	Tolmezzo	Farman	8
6ª Sq. Ric/Comb.	Chiasiellis	Farman	8
III Gruppo Sq.	**Verona, 1ª Armata**		
5ª Sq. Offesa	Verona	Caproni 300	5
1ª Sq. Ric/Comb.	Verona	Farman	7
12ª Sq. Ric/Comb.	Villaverla	Farman	5
3ª Sq. Caccia	Brescia	Aviatik	8
4ª Sq. Caccia	Verona	Aviatik	8
1ª Sq. Idro	Desenzano	FBA	9
Gruppo Aviazione per Artiglieria			
5ª Sq. Art.	Oleis 2ª Armata	Caudron	9
7ª Sq. Art.	Oleis 2ª Armata	Farman	8
1ª Sq. Art.	Risano 3ª Armata	Caudron	?
2ª Sq. Art.	Chiasiottis 3a Arm.	Caudron	10
3ª Sq. Art.	Gonars 3ª Armata	Caudron	9
4ª Sq. Art.	Gonars 3ª Armata	Caudron	9
6ª Sq. Art.	Verona 1ª Armata	Farman	?
Autonomous Squadriglie and Sezioni			
5ª Sq. Caccia	Milano	Aviatik	12
13ª Sq. Ric/Comb.	Valona	Farman	?
1ª Sq. Caccia	Campoformido	Nieuport	11
2ª Sq. Ric/Comb.	Campoformido	Farman	9
Sezione Difesa	Aviano	Farman	3
Sezione Difesa	Ancona	Farman	4
Sezione	Tripoli	Farman	5
Sezione	Bengasi	Farman	?

Val Lagarina and Val Sugana, and on 28 May Asiago fell. This dangerous attack met Italian resistance. Many forces were quickly shifted from the Isonzo to the Vicenza front, including aviation squadrons. In mid-June the enemy attack was stopped. The Italian counteroffensive struck back on 25 June, and after heavy fighting much lost ground was regained. The Austrians, however, had conquered a line of peaks, from Col Santo to Ortigara, that they turned into a formidable defence line from which they menaced the Italian plains.

In August 1916 the Italians attacked again with the 6th Battle of the Isonzo. Brilliant actions brought a breakthough of the enemy line, the conquest of the Mounts Sabotino, Oslavia, Podgora and San Michele, and finally the city

Above: The Carso is a hellish landscape of rocks and no water.

of Gorizia. High was the moral value of this victory, but strategically the situation did not change, the Austrian defence line simply moved back to another range of mountains, still formidable: Kuk, Vodice, Monte Santo, San Gabriele, Vertoiba and Faiti.

During autumn 1916, from 14 September to 4 November, the 7th, 8th and 9th Battles of the Isonzo took place, three short offensives called "*spallate*" (shoving pushes) with limited objectives, in the area between Gorizia and the sea, launched for international reasons, that is, to provide indirect help to a new ally, Rumania.

On 14 September 1916 two territorial air defence groups were formed, *Gruppo di Difesa antiaerea,* Northern and Souther, for the protection of the national territory. Nine days later air defence squadrons got new numbers, starting with 101. So, the Squadriglie that were in formation, 38^{a}, 39^{a} and 40^{a}, got the new numbers 101^{a}, 102^{a} and 105^{a}.

On 16 November the Command of the *Battaglione Squadriglie Aviatori* was discontinued, and its tasks concerning operations went to the new *Ufficio Servizi Aeronautici*, USA, of the Supreme Command, headed by *col.* Vittorio Buffa di Perrero while organisation tasks went to the *Ufficio Squadriglie* of the *Comando d'Aeronautica (Aviatori)* in Turin, is chief was Colonel Giuseppe Motta.

Up to this period, the Austrian *Luftfahrtruppe* and naval aviation had maintained their ascendancy, thanks to the better quality of their airplanes and flying boats (and to their daring and capable pilots), while on the Italian side only the Caproni bomber was a better weapon than those of the enemy. Italian fighters were all of French design, seaplanes depended on enemy technology and engines, and reconnaissance aircraft were French designs, obsolete Caudron, Voisin and Farman, but new designs were entering service, such as the SAML and Savoia Pomilio two-seaters and the Macchi L.3 seaplanes that were raising Italian aviation to a situation of parity. In 1916 the Italians claimed 60 air victories (besides the ones of the Navy and of the French). Of these, fighters claimed 31, reconnaissance airmen claimed seven, Caproni gunners twelve and anti-aircraft artillery twelve, with some double claims. Known Austrian losses were 42, with 21 prisoners, plus two deserters. The Austrians claimed about 55 victories, known Italian losses in air combat are about 30, with 23 dead and 18 prisoners.

On 19 January 1917 new squadron numbers were issued: from 100 to 199 for defence and colonial Squadriglie, from 200 to 220 for bomber Squadriglie of the Navy, from 221 to 250 for fighter Squadriglie of the Navy, from 251 onwards for seaplane Squadriglie.

About March 1917 *gen.* Leone Andrea Maggiorotti replaced *col.* Motta as chief of the Ufficio Servizi Aeronautici. On 5 April 1917 a new structure of the air force at war was approved: on 10 April the 2^{a} and 3^{a} Armata got their

Order of Battle 1 January 1917

IV Gruppo Pordenone (Comando Supremo)
1ª Caproni, Comina - cap. Toccolini
2ª Caproni, Aviano
3ª Caproni, Aviano - cap. Falchi
4ª Caproni, Aviano - ten. Molinari
6ª Caproni, Aviano - cap. Sibilla
7ª Caproni, Aviano - cap. De Muro
8ª Caproni, Comina - cap. Govi
10ª Caproni, Campoformido - cap. De Riso
13ª Caproni, Comina - cap. Oppizzi

III Gruppo Verona (Aeronautica 1ª Armata)
1ª Idro FBA, Desenzano - cap. Amour
5ª Caproni, Verona - cap. Rossi
9ª Caproni, Verona - cap. Reggio
31ª Farman, Verona - cap. Straziota
32ª Farman, Villaverla - cap. Zoboli
71ª Nieuport, Villaverla - cap. Chiaperotti
72ª Aviatik, Brescia - cap. Matteucci
73ª Nieuport, Verona - cap. Sanità
75ª Nieuport, Verona - cap. Gordesco

VII Gruppo Nove di Bassano (Aeronautica 1ª Armata)
46ª Farman, Castenedolo and Trizzino - cap. Adorni
49ª Caudron, Nove di Bassano - cap. De Sanctis

II Gruppo Udine (Aeronautica 2ª Armata)
27ª Farman, Campoformido - cap. Giorgi
29ª Farman, Cavazzo Carnico - cap. Ferrero
30ª Farman, Chiasiellis - cap. Aliquò Mazzei
35ª Voisin, Campoformido - cap. Carignani
76ª Nieuport, S. Maria la Longa - cap. De Carolis

VI Gruppo Oleis (Aeronautica 2ª Armata)
41ª Caudron, Oleis - cap. Martucci
45ª Farman, Oleis - cap. Giovine

I Gruppo S. Maria la Longa (Aeronautica 3ª Armata)
25ª Voisin, Pozzuolo del Friuli - cap. Stanzani
26ª Voisin, S. Maria la Longa - cap. Minellono
28ª Farman, Chiasiellis - cap. Neri
36ª Farman, S. Maria la Longa - cap. Clementi
77ª Nieuport, Cascina Farello - cap. Piccio

V Gruppo Chiasottis (Aeronautica 3ª Armata)
42ª Caudron, Medeuzza - cap. Negro
43ª Caudron, Bolzano - cap. Carranza
44ª Caudron, Gonars - cap. Beltrami

VIII Gruppo (Albania)
11ª Caproni, Tahyraqua - cap. Pascale
34ª Farman, Valona - cap. De Rada
Sezione Nieuport, Tahyraqua - ten. Sabelli

Autonomous Squadriglie
2ª Idro FBA, Grado - cap. Martellucci
12ª Caproni, Benghazi - ten. Raffaelli
33ª Farman, Campoformido - cap. Russi
37ª Farman, P.S. Pietro - cap. Oddo
47ª Farman, Saloniki - cap. Chiappelli
48ª Caudron, Belluno - cap. Porro
70ª Nieuport, S.Caterina - cap. Tacchini
74ª Aviatik, Trenno (Milan) - cap. Giorgi
78ª Nieuport, Istrana - cap. Bolognesi

Territorial Defence Units
101ª Farman, Bari - cap. Sisto
102ª Farman, Ancona - ten. Gallotti
103ª Farman, Brindisi - ten. Calleri
104ª Farman, Tripoli - cap. Suglia
105ª Voisin, Florence - cap. Rospigliosi
Sez. Difesa Foligno - ten. Debolini
Sez. Difesa Ravenna - ten. Cottino
Sez. Difesa Rimini - ten. Ponis
Sez. Difesa Comina - ten. De Gasperi
Sez. Difesa Benghazi - ten. Marchisio
Sez. Difesa Rome - ten. Masi
Sez. Difesa Bergamo - ten. Bourlot

Centro Formazione Squadriglie di Ghedi - cap. Ferraro
Centro Formazione Squadriglie di Arcade - cap. Degan

Comandi Aeronautici (aviation commands) that became a lean and effective instrument of joint coordination, and included control of the kite balloon units. Before that, the squadrons and groups only reported to the Army Corps. The *Comando Aeronautica* of the 2ª Armata received the II and VI Gruppo, with a total of nine Squadriglie, while the *Comando Aeronautica* of the 3ª Armata, with *t.col.* Lelio Gaviglio, got the I and V Gruppo. The III Gruppo was split, forming the IX Gruppo with Squadriglie 37ª (defence of Bergamo), 72ª, 73ª and 74ª (previously defence of Milan) and the new 75ª and 112ª. The IV Gruppo Caproni of the Comando Supremo formed the XI Gruppo with Squadriglie 3ª, 4ª, 6ª, 7ª and the new 14ª and 15ª. The X Gruppo was formed and assigned to the Supreme Command with Squadriglie 33ª, 70ª, 78ª, 79ª and 82ª.

Another new structure of the air force was approved shortly after, on 2 May 1917, that created the *Comando Aeronautica* of the 4ª and 6ª Armata and the new XII Gruppo. The 1ª Armata was assigned the III Gruppo, enlarged with the 46ª and the new 50ª, exchanging its 71ª with the 72ª of the IX Gruppo, that now included 37ª, 71ª, 73ª, 74ª, 75ª and 112ª. The 4ª Armata had the new XII Gruppo with 48ª, a flight of the 113ª, a flight of the 83ª Nieuport and the 35ª in formation. The 6ª Armata had VII

Above: General Giovanni Battista Marieni, an able technician, was chief of the General Directorate of Aviation in 1916 and 1917.

Above: The jacket of an Italian pilot, on the sleeve two stars indicate a *tenente*, and there is the crowned eagle of a military pilot and the insignia of a combat wound; on the chest there are the ribbons of two Silver Medals, a Croce di Guerra (war cross) and the medal for three years of war service. (Collection Catellani)

Gruppo reformed with 32ª, 49ª and 79ª from the X Gruppo and the new 26ª. The Command of *Squadriglie per Artiglieria* was abolished, and the units of its gruppi, V, VI and VII, lost the specification of *'squadriglie di artiglieria'*. Squadrons were then divided in "*d'Armata*", for strategic reconnaissance, and "*di Corpo d'Armata*" for tactical reconnaissance and artillery spotting, a rather virtual distinction, as they used the same airplanes.

In May 1917 *gen*. Cadorna launched the 10th Battle of the Isonzo, that brought the conquest of mounts Kuk and Vodice. It was one more massive effort, with terrible casualties on both sides that ended with irrelevant territorial gains. During the 10th Battle aviation participated massively, with up to 140 sorties in a single day, and many air combats developed, but the Austrians could not stop Italian formations from supporting the offensive.

From 10 to 25 June there was the bloody Battle of the Ortigara, the first Italian attack in the northern alpine sector. The Alpini attacked to reconquer the ground still held by the enemy after the "Strafexpedition" of 1916. The Ortigara Mountain was captured and then abandoned after terrible losses that were of no avail. Some squadrons were shifted from the Isonzo front to support this offensive, and air activity was intense. The Austrians suffered serious losses in the air. Italian fighter units now used Spad 7 and Nieuport 17 aircraft, which were better than the enemy fighters. The Italians were decidedly gaining supremacy in the air.

The offensive of the Bainsizza, the 11th Battle of the Isonzo, was fought from 18 August to 28 August 1917. It was a huge effort, massively supported by artillery, and with large-scale operations of the Italian air force. Combined formations of Italian aircraft gained total mastery of the air, despite the gallant efforts of Austrian fighters, a total of 1,474 war sorties were flown, on the 21st alone, 245 airplanes flew 338 combat sorties to support the attacking troops. Some twenty airplanes were lost, with a dozen dead airmen. The Bainsizza plateau was conquered, but the huge concentration of forces and the heavy Italian losses did not

Above: General Leone Andrea Maggiorotti, chief of the Ufficio Servizi Aeronautici.

Above: The prancing horse, the personal insignia of Francesco Baracca, here in a panel in the Francesco Baracca Museum in Lugo di Romagna.

bring the decisive victory that *gen*. Cadorna was hoping for. The Austrian army was on the brink of collapse, but the Italian attack failed short of the breakthrough, and the frontline simply moved back another range of hills.

On 17 September 1917 the *Comando d'Aeronautica* issued a new range of numbers for squadrons, both existing and future ones, according to their equipment: from 1 to 20 the Squadriglie using Caproni, from 21 to 40 the ones with SP.3, from 41 to 60 with SIA 7b1, from 61 to 69 with SP.4, from 70 to 85 Nieuport and HD, from 86 to 100 SVA and Spad, from 101 to 130 SAML-equipped and defence Squadriglie, from 131 to 150 Pomilio, from 151 to 160 SIA 7B2, from 161 to 170 SIA 9b, from 171 to 180 P 500 (what kind of aircraft was meant with P 500 is not known, probably a twin-engine Pomilio) from 181 to 189 Caproni triplane, from 241 onwards airplanes and seaplanes of the Regia Marina. This regulation considered a construction program that was never fully implemented: the P 500 never arrived, only one SIA 9B unit was formed, the 161ª, and only two Caproni Triplane units, the 181ª and 182ª. As a consequence of this order, the 73ª and 74ª SAML defence Squadriglie were re-named 121ª and 122ª.

While in other nations the propaganda exalted the exploits of fighter pilots, fueling the mythology of the aces, in the Italian air force fighter pilots shared the honors of the press with the "*capronisti*", whose exploits also earned Gold Medals, and also with the reconnaissance pilots and observers, commended for the number of reconnaissance flights carried out or photos taken.

In 1917, until 25 October, Austrian airmen claimed 185 victories, and the newly arrived Germans five. Available Italian documents report some 90 airplanes lost in air combat, plus 37 more shot down by artillery, with over 60 dead and 15 prisoners. Italian pilots claimed 120 victories, plus 12 for the artillery, while available Austrian data admit 105 losses in combat, plus five German planes. On 23 October *col.* Riccardo Moizo replaced Gen. Maggiorotti as chief of the *Ufficio Servizi Aeronautici*.

In the late summer and autumn 1917 the front stabilised and Italian troops settled for defence. A counteroffensive of the Austrian Empire was expected, considering that forces could be shifted from the Russian front. The commander of the 2ª Armata, *gen*. Luigi Capello, however, despite reports of the arrival of German reinforcements, ignored Cadorna's order to prepare for the defensive, believing that one more Italian attack was the better option.

Left: Three aces of the 91ª Squadriglia: Pier Ruggero Piccio, Francesco Baracca, and Fulco Ruffo di Calabria.

The Defeat of Caporetto

A German army had secretly reached the Italian front, and on 24 October 1917 the enemy under the command of German General Otto von Below attacked in an unexpected sector, at Caporetto (now Kobarid, Slovenia), quickly tearing through the lines of the 2ª Armata using new tactics, rushing down the valleys ignoring and leaving behind the strongholds on the hills. The Italian front collapsed, and moved back almost 100 miles, leaving behind a large number of prisoners, equipment and above all national territory. In those dramatic days, the Italian air units, that had failed to discover the mass-up of German troops, protected by bad weather and excellent camouflage, fought stubbornly. German fighter, bomber and reconnaissance squadrons (Jasta 1, 31 and 39, with Albatros D.III and D.V, FliegerAbteilung FA 2, 14, 17, 39 and FA Artillerie FAA 204, 219 and 232, and Reihenbildzug 1 with excellent Aviatik, DFW and Rumpler) had also reached the front, with eager and able airmen, but the Italians faced them effectively. Italian losses in air combat were very heavy, but German and Austrian losses were just as heavy. 22 airfields had to be abandoned quickly, a large number of aircraft and equipment was lost, and operative conditions became terrible, but all Squadriglie, while retreating, never stopped opposing the enemy advance. Their role was very important: they strafed and bombed the enemy, dropped supplies to the troops, and showed to the soldiers that the sky was still contested.

Reconnaissance squadrons concentrated first in the area of Ghedi and Bergamo, then they were moved to Veneto, while at Ghedi bombers were concentrated. The retreat caused a very bad crisis in all units: most reconnaissance squadrons could fly just one or two airplanes, instead of nine. The crisis was made worse by the losses of large quantities of aeronautical equipment that could not be replaced quickly as new types (Hanriot, SIA 7A, Pomilio P) were just entering production. On account of the dramatic situation, all Squadriglie equipped with Savoia Pomilio, gathered at Ghedi and Ponte S. Pietro, were rushed back to the front: Squadriglie 22ª, 27ª, 35ª and 36ª to Istrana, assigned to the XII Gruppo, 23ª and 28ª to Bazzera, 38ª and 39ª to Marcon, to the V Gruppo.

The Italian Government on 1 November 1917 created the *Commissariato Generale d'Aeronautica*, within the Ministry of Weapons and Munitions, with the task of co-ordinating and developing industrial mobilisation and production of aircraft and engines, that replaced the *Direzione Generale d'Aeronautica*. Eugenio Chiesa was appointed as its chief: he was a Republican member of Parliament, part of the radical left, a Freemason, an old man of little specific experience but of strong authority and high moral standing. The *Commissariato* was to all effects a Ministry, but it did not take that name so that Chiesa didn't have to swear fealty to the King.

On 8 November 1917 the *Ufficio Servizi Aeronautici* disbanded the Comandi d'Aeronautica of the 2ª and 6ª Armata, and reformed the Command of the 4ª Armata, basically with the same units of the 2ª, and formed a *Comando Truppe degli Altipiani* (Command for the Troops of the Asiago Plateau). Two new groups were formed, XIII and XIV Gruppo, three reconnaissance squadrons were disbanded, 25ª, 44ª and 50ª, while four more, 21ª, 24ª, 40ª and 62ª (the last one never used in operations) were moved away from the front to re-equip.

Field Marshal Conrad, on 9 November, the day in which Cadorna was replaced as supreme commander by Armando Diaz, launched an attack on the front of the Asiago plateau,

Above: Caporetto, now Kobarid, and the difficult terrain surrounding it. (Jost Gantar)

joining up with the right wing of von Below's 14th Army which unsuccessfully attacked the defenses on Mount Grappa. At the same time the Austrians also attacked to the south, establishing a bridgehead beyond the Piave at Zenson, but at Cortellazzo their march towards Venice was stopped, despite the support from the sea from the battleships *Wien* and *Budapest*. It was the crucial moment of the conflict, the Allies, who had sent six French and three British divisions to reinforce Italy, with many fighter and reconnaissance squadrons, advised a retreat to a new line on the Mincio River, and kept their forces in the rear, fearing the worst (they would enter the line only on 4 December, when things were settled), but the Italian Army was able to resist. The Austrians attacked on the highlands, until 26 November, and also on the Piave River, exploiting the bridgehead at Zenson, but they were stopped. In December the fighting continued, the Austrians attacked on Sisemol, and on 5 December they conquered the mounts Melette di Foza and Gallio and finally, on 23 December, Colle Eckar, Mount Valbella and Col del Rosso, the Germans conquered Col Caprile and Asolone, but the line of defense on Mount Grappa held, and that resistance determined the outcome of the war.

On 20 November 1917 the Italian air force had a strength of 59 Squadriglie and 2 sezioni with 378 aircraft (including 59 Caproni, 9 Farman, 5 Caudron, 59 SAML 200, 1 SP.2, 55 SP.3, 19 SP.4, 28 Pomilio, 6 SIA 7B, 31 Nieuport, 36 Spad 140, 34 Hd.1, 36 various) and 497 pilots (258 officers and 239 enlisted men) 284 observer officers, 152 gunners. On that date, the Navy had 119 pilots plus 79 Army pilots assigned to its units.

On 22 November the Inspectorate of Fighter Squadrons, *Ispettorato Squadriglie da caccia* was created, entrusted to the ace *t.col.* Piccio. In December the *Centri Formazione Squadriglie* (centers for the formation of new squadrons) moved from Arcade, San Pelagio and Ghedi, now too close to the frontline, to Busto Arsizio, Riva di Chieri and Ponte San Pietro.

On 4 December 1917 the Austrians attacked once more in the Trentino, trying to break through to the plateaus. Initially they were successful and gained ground, but soon they were stopped, just as their final attacks along the Piave River and the Mount Grappa, which marked the new Italian line of resistance, were also checked. The air force was heavily engaged in the combats of December, with its few and obsolete aircraft, and Austrian fighters scored their last meaningful victories. On 7 December 150 airplanes of III and IX Gruppo attacked on the Asiago Plateau, not all went well, coordination was deficient and Austrian fighters shot down five planes, but it was the demonstration that Italian aviation was back in force.

The turning point in the struggle for air superiority was the so-called Battle of Istrana, 26 December 1917, when the German air force attacked Istrana airfield and was faced by the local fighters, that shot down eleven German bombers

Above: Ten. Amerigo Contini, a Caproni pilot and an artist, drew very realistic paintings of bombers in action, that were used for postcards.

with no loss to themselves. On 21 December the instruction came to paint roundels on the wings of airplanes, like the Allies, and as the Navy had been doing since 1916, but as it did not specify the order, there were roundels both with red outside and green inside, and the opposite.

The Events of 1918

The first part of 1918 was quiet all along the front. On 28 January there was the successful attack of Col del Rosso, the first Italian return to the offensive. The Italian Army was reorganized and strengthened under the leadership of General Armando Diaz, who had replaced Cadorna. Italian industries were now producing plentiful weapons, and fighter units received a large quantity of the well-liked Hanriot Hd.1 and the Macchi M.5 seaplane, while France supplied Nieuport 27s and Spad 13s. The new reconnaissance SIA 7B and Pomilio PE models and the Caproni 600, or Ca.5 bomber were also mass produced, but they turned out to be most troublesome in service, so that reconnaissance and bomber units faced a very serious crisis. In order to compensate for the arrival of French fighter and reconnaissance units on the Italian front, a Caproni Group with three Squadriglie was sent to the French front. On 1 February 1918 the Comando d'Aeronautica of the 4ª Armata was reformed, commanded by *col.* Zanuso.

A new airplane of Italian design reached the front, the SVA. Unfit for fighter use, it was the ideal fast reconnaissance machine. On 22 February the Command decided to activate six Sezioni SVA of six aircraft each to be assigned to each Army for strategic reconnaissance. The three already existing ones were 1ª Sezione assigned to the 1ª Armata, at Sovizzo, the 2ª Sezione to the *Truppe degli Altipiani* at Nove, and the 3ª Sezione to the 4ª Armata, at Isola di Carturo. Each Sezione SVA was assigned to a Squadriglia for support in administration. Two other squadrons were to be formed with SVAs, the first of which was the crack 87ª Squadriglia.

On 1 March the *Ufficio Servizi Aeronautici* was disbanded, and the *Comando Superiore di Aeronautica* replaced it. Its commander was *gen.* Luigi Bongiovanni, a capable and intelligent officer. On 17 March 1918 the assignement of units to the Army commands was changed once more in order to conform the new organization to the presence of Allied units. The *Comando d'Aeronautica* of the 2ª Armata was reformed, the command of the 7ª Armata was created too, while the command of the 6ª Armata was scaled down to Aeronautics Bureau.

The development program brought the activation of new Gruppi, leaner commands with squadrons with the same type of airplanes. The program considered fighter Squadriglie of six flights, with a total of 36 airplanes and pilots. The *Squadriglie di Corpo d'Armata* (army corps squadrons, tactical reconnaissance) were to have 14 aircraft and crews, flying the new SIA 7B2. The units to be so equipped were 22ª, 23ª,

Order of Battle May 1917

Aeronautica del Comando Supremo
IV Gruppo
Squadriglie 1ª, 8ª, 10ª, 13ª, 14ª
X Gruppo
Squadriglie 33ª, 70ª, 78ª, 79ª, 91ª
XI Gruppo
Squadriglie 2ª, 3ª, 4ª, 6ª, 7ª, 15ª
Comando d'Aeronautica 1ª Armata
III Gruppo
Squadriglie 5ª, 9ª, 31ª, 46ª, 72ª, 1ª Idro
IX Gruppo
Squadriglie 37ª, 71ª, 73ª, 74ª, 75ª
Comando d'Aeronautica 2ª Armata
II Gruppo
Squadriglie 21ª, 22ª, 27ª, 30ª, 40ª, 41ª, 76ª, 81ª
VI Gruppo
Squadriglie 24ª, 29ª, 45ª, Sez/113ª
Comando d'Aeronautica 3ª Armata
I Gruppo
Squadriglie 23ª, 25ª, 36ª, 77ª, 80ª, 2ª Idro
V Gruppo
Squadriglie 38ª, 39ª, 43ª, 44ª
Comando d'Aeronautica 4ª Armata
VII Gruppo
Squadriglie 32ª, 49ª, 79ª
Comando d'Aeronautica 6ª Armata
XII Gruppo
Squadriglie 48ª, Sez/113ª, Sez. Caccia
Albania
VIII Gruppo
Squadriglie 11ª, 34ª, Sez/85ª
Macedonia
Squadriglia 47ª, Sez/83ª
Colonies
Squadriglie 12ª, 104ª, Sez. Tripoli, Sez. Tobruk
Squadriglie da Difesa
101ª, 102ª, 103ª, 105ª and 14 Sezioni

26ª, 27ª, 28ª, 31ª, 32ª, 33ª, 35ª, 36ª, 37ª, 38ª, 39ª, 48ª, 61ª, 62ª, 114ª, 118ª, 121ª, and overseas 111ª and 116ª. Twelve Pomilio type E or type F were to equip *Squadriglie da ricognizione d'armata* (army, or strategic, reconnaissance) converting the 113ª, 117ª and 120ª. Eight Caproni 600 or Caproni triplanes had to equip bomber squadriglie. This program was wrecked by the failure of the SIA 7B, but it was too ambitious anyway, and at the end of the war fighter squadrons had at best 18 aircraft, not 36. As for the Caproni 5, or Ca 600 hp, a huge construction program was started. Factories were built at Rome (Bastianelli) Pistoia (Breda Pistoia) Reggio Emilia (later Officine Reggiane) Milan (Breda) besides Caproni in Milan. This bomber however never took off. The first ones were tested at the 4ª Squadriglia and sent to the Squadra San Marco, an élite Navy unit based in Venice and commanded

Order of Battle 1 August 1917

Comando Supremo
Raggruppamento Squadriglie da Bombardamento, Pordenone
IV Gruppo
Squadriglie 1ª, 8ª, 10ª, 13ª, 14ª Caproni
XI Gruppo
Squadriglie 2ª, 3ª, 4ª, 6ª, 7ª, 15ª Caproni
X Gruppo
Squadriglie 70ª and 82ª Nieuport, 91ª Spad S.Caterina, 114ª SP Campoformido
Comando Aeronautica 1ª Armata
III Gruppo, Verona
Squadriglie 5ª and 9ª Caproni, Verona, 31ª SP Castelgomberto, 46ª MF Castenedolo, 50ª MF Villaverla, 72ª SAML Brescia, 1ª Idro FBA Desenzano
IX Gruppo
Squadriglie 37ª SP2 and SP3 Ponte S Pietro, 71ª Nieuport Villaverla, 73ª SAML and 75ª Nieuport Verona, 74ª MF/ SP Trenno
Comando Aeronautica 2ª Armata
II Gruppo
Squadriglie 21ª, 22ª, 33ª, 35ª and 40ª SP, 132ª P Campoformido, 27ª SP Chiasiellis
VI Gruppo
Squadriglie 24ª SP, 29ª MF, 113ª SAML and III/83ª Nieuport Cavazzo Carnico, 41ª SP and 45ª MF Oleis, 76ª and 78ª Nieuport-Spad and 81ª Borgnano
Comando Aeronautica 3ª Armata
I Gruppo
Squadriglie 23ª SP and 84ª Nieuport S. Maria la Longa, 25ª Voisin Pozzuolo, 77ª Nieuport Spad and 80ª Nieuport Aiello, 112ª SAML Lavariano, 131ª P Chiasiellis, 2ª Idro FBA Grado
V Grupppo
Squadriglie 28ª SP S.M.la Longa, 38ª SP Risano, 39ª SP Sammardenchia, 42ª Caudron Medeuzza, 43ª Caudron Bolzano, 44ª Caudron Gonars
Comando Aeronautica 4ª Armata
XII Gruppo, Belluno
Squadriglie 48ª Caudron and II/83ª Nieuport Belluno, I/113ª SAML Villaverla
Comando Aeronautica 6ª Armata
VII Gruppo
Squadriglie 26ª SP Casoni, 32ª Farman S. Pietro in Gu, 49ª Caudron Nove, 79ª Nieuport Istrana
Albania
VIII Gruppo, Valona
Squadriglie 11ª Caproni, 34ª Farman, Sezione Nieuport
Macedonia
Squadriglie 47ª Farman, I/83ª Nieuport

Above: Kaiser Wilhelm II of Germany visited the Italian front: he is here on the Castle of Udine, the city where the Italian supreme command was, before the retreat in October 1917.

by the poet Gabriele d'Annunzio but they could never be used operationally, because they had never-ending teething troubles, particularly with their Fiat A12 engines. Many were lost in fatal accidents.

Another batch was supplied to the US Navy, for its Northern Bomber Group, but they also were a failure, and many were lost during the ferry flight to France. At the end of the war, there were hundreds of partially completed Caproni 5 in Italian factories, but this bomber was actively serving only with 6ª Squadriglia, while the ones of San Marco unit of the Navy were useless. After the war, the air force didn't know what to do with them: it was proposed to replace the Fiat engines with less powerful (but more reliable) SPA 6A engines and use them as colonial bombers. Some were turned into passenger transports, again with tragic results, as one of them, with two pilots and 15 journalists, crashed in Verona in August 1919, it was the worst air accident so far. One journalist who was invited for that trip, but missed the airplane, was Benito Mussolini.

The situation was so bad that the old and faithful Ca.3 450 hp was to be re-introduced in production. The Ca.3 actually remained in active service with bomber squadrons as late as 1927, and it was built in large quantities even in the early Twenties.

On 25 May 1918 the *Comando d'Aeronautica a Disposizione*, (command of available aviation) was formed replacing the *Comando d'Aeronautica del Comando Supremo*. On 12 June the 8ª Armata got its own *Comando d'Aeronautica*, while the one of the 2ª Armata was disbanded. The development programme issued on 17 March had to be scaled down: squadriglie 121ª, 39ª and 118ª kept their old SAMLs instead of converting on the Pomilio (actually a lucky event for them) while the 23ª, 26ª, 48ª, 61ª, 62ª and 114ª converted on the Pomilio instead of the SIA. But as the number of accidents of Pomilios and even more of SIAs kept growing, bringing some units on the verge of mutiny, the *Comando Superiore di Aeronautica* ordered all surviving SP.3, SP.4 and Caudron G.4 to be kept in service. An ever more important mission was the launch of propaganda leaflets in German, Czech, Hungarian, Polish, that a special office prepared to stoke the nationalism of the peoples that Vienna ruled.

On 5 June, expecting a new Austrian offensive (that started on 15 June, and was called the Battle of the Piave) a *Massa di Manovra Squadriglie da caccia* was formed, commanded by *t.col.* Piccio, with squadrons 70ª, 82ª, 91ª, 75ª, 72ª, 76ª, 79ª and 80ª, usually called *Massa da Caccia*, Fighter Mass. While the Austro-German offensive of Caporetto, 24 October 1917, predicted only in vague terms, actually had surprised the Italians, the Austrian strategy of June 1918 was controlled by the Italian intelligence services, and it was monitored by the reconnaissance missions. When on 15 June the enemy attacked, all along the front, from the mountains to the sea (an error due to the rival strategies within the Austrian command, which precluded a single strong attack) the Italian defenses were ready and well placed, and the offensive was stopped on June 21, putting to all effects the Austrians out of the war, and indirectly destroying the remaining hope of victory of Germany.

At the time of the Battle of the Piave the Italian air force at the front included 65 squadrons and 9 flights, with 647 aircraft (three fourths of them serviceable) and 770 pilots, 474 observers, 176 gunners, 916 mechanics and 477 fitters.

The Italian victory in the air during the battle of the Piave was staggering. The Italian airmen gained absolute superiority. The two-seaters and bombers could intervene in the crucial sector, strafing and bombing the bridges and the barges that the Austrians were using to get as many forces as possible across the Piave River. The Austrian offensive, although valiant, was defeated. It was poorly thought out, with attacks on the whole front, without the choice of a precise objective, it was from the start destined for defeat, unless one condition had occurred, the moral collapse of the Italian soldiers. The intervention of the air force shortened it by several days, indirectly saving the lives of countless soldiers on both sides. The allies claimed 107 victories during the battle, the Austrians less than half, but Italian known losses amounted to eighteen aircraft, almost all fighters, and among them the Spad of Francesco Baracca, Italy's top ace, three balloons and six RAF Sopwith Camels.

After this victory, the Fighter Mass was disbanded. The Aviation Command issued a plan to bring up all reconnaissance squadrons to a force of 12 aircraft and 14 pilots and 14 observers. According to this plan, most SAML, SP.3 and SP.4 squadrons had to convert on the Pomilio, while the other SP.3 squadrons had to convert on the SIA 7B. This plan, however, failed because after a series of accidents, all the SIAs had to be retired and grounded, while the Pomilio itself was feared and disliked. Instead of increasing, the Italian reconnaissance force was drastically cut. The following squadrons were disbanded, dividing their pilots among surviving units: 24ª, 35ª, 37ª, 40ª, 62ª, 117ª, 132ª, 135ª and 139ª.

After the Piave victory, the Austrian air force almost disappeared from the skies, having suffered terrible losses. It was not a spent force, however, and during the final offensive of October 1918 it was again in the air, fighting with valor and still inflicting hard losses to the Italians, but again suffering much heavier losses itself.

August 1918 was marked by the feat of the Flight to Vienna of the poet Gabriele d'Annunzio with the 87ª Squadriglia SVA: taking off from San Pelagio, nine SVA of this crack unit reached Vienna and dropped leaflets. This flight was publicized at the time, and even more in later years, and it was indeed a brilliant feat. The presence of d'Annunzio flying as a passenger over the center of the enemy capital was useful as a proof that the Italian planes could have carried bombs, at least equal to the weight of the poet, so the dropping of leaflets only was a deliberate choice of a non-lethal demonstration of power.

The Regia Marina also tried to organize its own strategic land bombers. Its first squadron was 201ª Squadriglia,

Above: Col. Moizo, second from the left, a pioneer pilot in Libya in 1911, was chief of the Ufficio Servizi Aeronautici and of the Italian air force post-war. His career peaked with the command of the Carabinieri military police.

active in 1917, then the Navy was to receive 181ª and 182ª Squadriglia with Caproni triplanes (another bad airplane) but these were used by the Army that returned them only in October 1918. Plans were drawn to form huge fleets of bombers, based both in the North and in Apulia, to wreck the Adriatic bases where the Austrian "fleet in being" was moored. A large number of American pilots were to man these Capronis. A first effort was made on 17 July 1918, with a mass bombing of Pola. It was a brilliant action, with a coordination of dozens of Capronis, SIA 9B single-engine bombers of the d'Annunzio San Marco squadron, fighters, naval seaplanes and vessels etc. It is amazing how Italy could realize an excellent interforce cooperation in WW1, and no cooperation at all in WW2.

Another very important sector was home defense. There were many dozen airplanes wasted in the protection of home cities (Milan, Bologna, Florence, Rome, Naples etc.) that never went into action. A single mission of the German Navy Zeppelin LZ 59 over Naples on March 10, 1918 was enough to force Italy to waste a large number of men, guns and aircraft for home protection.

Italy had a large number of airships, both of the Army and Navy, of the Italian semi-rigid type. Their war actions are a valiant history, that still has to be told for foreign readers. Suffice to say that Italian airships of both services were in action, bombing the enemy, from the first day of war, 25 May 1915, to the last one, 4 November 1918.

In the period from July 1918 to the eve of the final offensive, the Italian fighters claimed about 85 victories, reported in the daily bulletins but not all approved in the controls after the war, of these thirteen went to the 91ª Squadriglia, a unit which after Baracca's death seems to have lost a little of the previous vehemence. Of these claims, about 40 correspond to known Austrian losses. Eleven victories were credited to flak, at least five of which were confirmed. The Royal Air Force, which applied more aggressive tactics

Order of Battle 1 October 1917

Comando Supremo
Raggruppamento Squadr. da Bombardamento, Pordenone, magg. La Polla
IV Gruppo - La Comina, magg. Bonazzi
Squadriglie 1ª, 8ª, 13ª Comina, 10ª and 14ª Campoformido
XI Gruppo - Aviano, magg. Armani
Squadriglie 2ª, 3ª, 4ª, 6ª, 7ª Aviano, 15ª Campoformido
X Gruppo - Udine, magg. Piccio
Squadriglie 70ª and 82ª Nieuport, 91ª Spad S. Caterina, 78ª Nieuport Istrana, 132ª Pomilio
Campoformido
Comando Aeronautica 1ª Armata
III Gruppo - Verona, magg. Resio
Squadriglie 5ª and 9ª Caproni Campoformido, 31ª SP Castelgomberto, 50ª MF Villaverla, 120ª
SAML Castenedolo, 1ª Idro FBA Desenzano, 3ª Idro FBA Pilzone
IX Gruppo - Villaverla, magg. Melegari
Squadriglie 37ª SP2 and SP3 Castenedolo, 71ª Ni/Spad Villaverla, 75ª Ni/Spad and 121ª SAML
Verona, 122ª Saml/SP Trenno
Truppe degli Altipiani (Troops of the Asiago plateau)
VII Gruppo - Nove di Bassano, cap. de Sanctis
Squadriglie 26ª SP.3 and 62ª SP.4 Casoni, 32ª MF S. Pietro in Gu, 33ª SP.3 and II/115ª SAML
Nove, 79ª Nieuport Istrana
Comando Aeronautica 2ª Armata
II Gruppo - Udine, magg. Zanuso
Squadriglie 21ª, 22ª and 40ª SP, 114ª SAML, 133ª P Campoformido, 27ª SP Chiasiellis (assigned to the VI Gr.)
VI Gruppo - Oleis, magg. Bongiovanni
Squadriglie 24ª SP, 113ª SAML and III/83ª Nieuport Cavazzo Carnico, 36ª SP and 45ª Farman
(disbanding) Oleis, 76ª Ni/Hd/Spad and 81ª Spad Borgnano
Comando Aeronautica 3ª Armata
I Gruppo - S. Maria la Longa, t.col. Novellis
Squadriglie 23ª SP S. Maria la Longa, 25ª Voisin Pozzuolo, 77ª Ni/Spad and 80ª Nieuport Aiello,
112ª SAML and 131ª P Lavariano, 2ª Idro FBA Grado
V Grupppo - Chiasottis, t.col. Santi
Squadriglie 28ª SP S. Maria la Longa, 38ª SP Risano, 39ª SP Sammardenchia, 42ª and 43ª Caudron
Medeuzza, 44ª Caudron Gonars
4ª Armata
XII Gruppo - Belluno, cap. Porro
Squadriglie 35ª SP.3 S Giustina, 48ª Caudron and II/83ª Nieuport Belluno, I/115ª SAML
Villaverla
Albania
VIII Gruppo - Valona, magg. De Rada
Squadriglie 11ª Caproni and Sezione Nieuport Tahyraqua, 34ª MF Valona
Macedonia
Squadriglia 47ª MF and I/83ª Nieuport Saloniki

Order of Battle 20 November 1917

Comando Supremo
X Gruppo - magg. Matteucci
Squadriglie 91ª Padova, 70ª, 82ª Istrana
Raggruppamento squadriglie da bombardamento
XI Gruppo - magg. Armani
Squadriglie 4ª, 5ª, 6ª Padova
IV Gruppo - magg. Bonazzi
Squadriglie 1ª, 8ª, 9ª, 13ª San Pelagio
XIV Gruppo - magg. De Muro
Squadriglie 2ª, 3ª, 7ª, 10ª, 14ª, 15ª Ghedi

Aeronautica della 1ª Armata - col. Gilbert de Winckels - Verona
III Gruppo - ten. col.Resio - Verona
Squadriglie 37ª Brescia, 72ª, 120ª, 1ª Sez./134ª Castenedolo, 75ª, 121ª Verona
IX Gruppo - magg. Melegari - Villaverla
Squadriglie 61ª, 71ª, 2ª Sez./134ª Villaverla, 31ª Castelgomberto

Aeronautica Truppe degli Altipiani - ten. col.De Masellis - Bassano
VII Gruppo - ten. col.Visconti - Nove
Squadriglie 33ª, 2ª Sez./139ª, 115ª Nove, 26ª, 32ª S.Pietro in Gu, 79ª Nove

Aeronautica della 4ª Armata - col. Prandoni - Castelfranco
II Gruppo - ten. col.Zanuso - Casoni
Squadriglie 113ª, 114ª, 132ª, 133ª Casoni
XII Gruppo - cap. Porro - Cittadella
Squadriglie 27ª, 48ª, 35ª Castello di Godego, 22ª, 36ª Istrana
VI Gruppo - magg. Calori - Istrana
Squadriglie 76ª, 78ª, 81 Istrana

Aeronautica della 3ª Armata - col. Novellis
I Gruppo - cap. Bevagna
Squadriglie 117ª, 131ª, 112ª Padova
V Grupppo - ten. col.Santi - Carpenedo
Squadriglie 23ª, 28ª Marcon, 38ª, 39ª Ca'Tessera
XIII Gruppo - cap. Gordesco - Marcon
Squadriglie 77ª, 80ª, 84ª, 2ª Sez./83ª Marcon

such as "offensive patrols", was less constrained in escort missions and had more ease in writing combat reports, credited itself with around 120 victories, which can be combined with around 50 known Austrian losses.

After the battle of the Piave, the Italian Air Force had acquired a superiority that allowed it to maintain dominance despite the gaps opened in its lines by the events of the SIA 7B and the Ca.5. The organization was now fully mature, from a logistical point of view, with large parks in Taliedo, Rubiera, Poggio Renatico and Padua into which the new airplanes from the factories flowed and were then distributed to the units. From 21 August to 20 September 1,154 airplanes were delivered, in the following month another 699, of which 302 were sent to the front, the others to schools, 320, and to the Navy.

The Final Offensive

On 21 October 1918 General Bongiovanni, chief of the air force, gave the orders for the final battle, long delayed and much pushed for by the Allies, who were in the offensive since August. The Italian offensive started on 24 October and was called the Battle of Vittorio Veneto. The *Massa da Caccia* was reformed using Squadriglie 91ª, 71ª and 72ª at Quinto, 70ª and 82ª at Gazzo, 77ª at Marcon, with a British squadron. A *Massa da Bombardamento* (bomber mass) was also formed with the IV, XI and XIV Gruppi with nine Caproni squadrons, the new XXII Gruppo with two SVA squadrons, and the Navy's two squadrons of Ca.4 triplanes. There were also the Squadra San Marco, the new 161ª Squadriglia SIA 9B and two Ca.5 squadrons with mostly American personnel, 203ª and 204ª. Two new aviation

Above: Supplied by France and license-built in Italy, the Hanriot Hd.1 was successful and much liked. This is Hd.1 19309 painted as French 515 with the colors of ace Baracchini in the Italian Air Force Museum at Vigna di Valle, near Rome.

commands were formed for the new 10ª and 12ª Army, while strategic reconnaissance was tasked to the 87ª Squadriglia and the new Sezione per la Cavalleria.

At the end of the war Italy fielded 25 groups, 64 squadrons and 6 flights at the front, with 1.055 combat ready aircraft, plus 20 defense squadrons. The Navy had 103 airplanes and 241 seaplanes at the Adriatic front, and 5 airplanes and 140 seaplanes in the coastal defense units.

The capacity of the industry, and of the training establishments, was such that this force could have grown very much, had the war lasted into 1919. But there were many problems that surfaced during the war, especially deficiencies in Italy's industrial basis, that were not studied after the war. All the positive lessons of the war (interforce cooperation of Army and Navy, industrial planning with the Allies, concentration of forces for strategic initiatives) were completely disregarded in the following years, leading to the defeat of the armed forces organized by Fascism during World War 2.

On 4 November 1918 the armistice took effect. Italy had won, and its aviators had given an important contribution, leaving a heritage of valour, from the top ace Francesco Baracca to the countless forgotten pilots, observers, gunners and mechanics, and paying the price of almost 1,000 casualties, one third due to enemy action, two thirds to accidents at the front or in training. The Air Force counted 7,758 pilots in the Great War, but a good part of them did not arrive at the front and were still students at the armistice. Considering perhaps a thousand observers and a few hundred machine gunners, we can assume that 4,500 men fought on the sky front in the Great War. Twenty-two aviators were decorated with the Gold Medal, 46 with the Military Order of Savoy, 1,233 with the Silver Medal and 820 with the Bronze Medal. Of them about a thousand fell, the losses for Army Aviation were 873, of which 205 in airborne war actions, 603 by accidents and the others for various or unknown causes.

On 10 November 1918 the Comando Superiore di Aeronautica became the Comando Generale d'Aeronautica. Disbandment of units began, while squadrons were prepared to move to Lybia. A period of decline began, and when in 1923 the independent air force, the Regia Aeronautica, was

Facing Page, Below: Eugenio Chiesa, Commissioner for Aeronautics in 1918, here during an event at the Ansaldo factory in Genoa.

Above: The Melette, mountains north of Venice, where harsh combats took place in the winter 1917/1918. This photo was taken during the summer.

Above: General Luigi Bongiovanni, the capable chief of the Italian air force in the war, with two officers of his staff: col. Amedeo de Siebert, chief of the Air Force in 1922–1923, and col. Franchini Stappo.

Above: This SP3, 4648, was shot down on 16 July 1917 by the Austrian pilot Kadetz of Flik 23/D, but research in the Italian archives has failed to find details on this event, proving that confirmation of legitimate claims is not always possible.

Right: The Italian "Medaglia al Valor Militare", medal for military valor, was assigned as Oro, Argento, and Bronzo; that is gold, silver, and bronze.

Below: The most famous airplane of the most famous deed of the Italian Air Force in World War 1, the SVA two-seater of Palli and d'Annunzio that flew over Vienna, is preserved in the Vittoriale, d'Annunzio's mansion, now a museum and research center, on Lake Garda.

Order of Battle Following the Changes of 17 March 1918

Comando Aeronautica 1ª Armata - col. Gilbert de Winckels
III Gruppo Ricognizione d'Armata e Caccia, t.col. Resio
134ª Pomilio, 75ª Nieuport, 1ª Sezione SVA -Verona, 135ª Pomilio, 71ª Spad - Sovizzo

XVI Gruppo Squadriglie di Corpo d'Armata
31ª SP.3, 61ª SP.4, 121ª SAML - Sovizzo

Comando Aeronautica 2ª Armata - t.col. De Masellis
XV Gruppo Ricognizione d'Armata e Caccia, t.col. Visconti
78ª Hanriot, 79ª Nieuport, 115ª SAML, 139ª Pomilio - S. Luca
XIX Gruppo Squadriglie di Corpo d'Armata, cap. Mutti
114ª SAML, 118ª SAML, 3ª Sez./24ª - Istrana

Comando Aeronautica 3ª Armata - t.col. Novellis di Coarazze
I Gruppo Ricognizione d'Armata, cap. Sella
117ª SAML, 131ª Pomilio - Fossalunga
V Gruppo Squadriglie di Corpo d'Armata, cap. Messi
28ª SIA, 62ª SP.4, 1ª Sez./24ª Marcon, 38ª SIA, 39ª SP.3 - Ca'Tessera

XIII Gruppo Caccia, cap. Gordesco
77ª Spad, 80ª Hanriot - Marcon

Comando Aeronautica 4ª Armata - col. Gallina
II Gruppo Ricognizione d'Armata, t.col. Zanuso
132ª Pomilio, 133ª Pomilio, 3ª Sez. SVA - Isola di Carturo
VI Gruppo Caccia, magg. Calori
76ª Hanriot, 81ª Nieuport - Casoni
XII Gruppo Squadriglie di Corpo d'Armata, cap. Martucci
22ª SIA, 35ª SIA, 36ª SIA, 2ª Sez./24ª - Casoni
27ª SIA, 48ª Cau. G.4 - Castello di Godego

Aeronautica 5ª Armata - col. Prandoni
113ª SAML, 23ª SP.3 - Medole

Ufficio Aeronautica 6ª Armata - magg. Beltramo
VII Gruppo Sq. di Corpo d'Armata e Caccia, cap. de Sanctis
26ª SP.3, 32ª SIA, 33ª SP.3, 1ª Sez./83ª - S. Pietro in Gu

Aeronautica 7ª Armata - col. Prandoni
IX Gruppo Ricognizione d'Armata e Caccia, magg. Chiaperotti
72ª Hanriot and Sez. Farman, 2ª Sez./83ª, 120ª SAML, 2ª Sez. SVA - Castenedolo
XX Gruppo Squadriglie di Corpo d'Armata, cap. Lanciani
37ª SP.3, 40ª SIA - Castenedolo

Assigned to the Comando Supremo - t.col. La Polla
Raggruppamento Squadriglie da Bombardamento, magg. Armani
IV Gruppo, magg. Bonazzi
1ª Caproni, 5ª Caproni, 8ª Caproni, 13ª Caproni - San Pelagio
XI Gruppo, cap. Nardi
4ª Caproni, 6ª Caproni Verona, 181ª Ca.4 - Ghedi
XIV Gruppo, cap. De Renzis
2ª Caproni, 7ª Caproni, 9ª Caproni, 10ª Caproni, 16ª Caproni - Padova
X Gruppo Caccia, magg. Matteucci
70ª Hanriot, 82ª Hanriot – Istrana, 91ª Spad – Quinto, 87ª SVA - Ghedi

Order of Battle October 1918

Comando Aeronautica a Disposizione – Padova
- 87a Squadriglia SVA – San Pelagio
- IV Gruppo – 1a, 5a, 8a, 13a Squadriglia Caproni – San Pelagio
- IX Gruppo – 4a, 6a Squadriglia Caproni – Ca' degli Oppi
- XIV Gruppo – 2a, 7a, 10a Squadriglia Caproni – Arquà Petrarca
- XXII Gruppo – 89a e 90a Squadriglia SVA – Busiago
- X Gruppo – 70a , 82a Squadriglia – Gazzo, 91a Squadriglia - Quinto di TV

Comando Aeronautica I Armata – S. Martino Buonalbergo
- III Gruppo – 61a, 134a, 75a Squadriglia, 1a Sezione SVA – Ganfardine
- XVI Gruppo – 31a, 71a, 121a Squadriglia – Castelgomberto
- 1a Squadriglia Idrovolanti – Desenzano

Comando Aeronautica III Armata – Carpenedo
- I Gruppo – 131a Squadriglia, 5a Sezione SVA – Marcon
- V Gruppo – 28a, 38a, 39a Squadriglia – Malcontenta
- XIII Gruppo – 77a, 80a Squadriglia – Marcon
- Gruppo Speciale I – 9a Squadriglia, 2 Sezioni – Ca'Tessera

Comando Aeronautica IV Armata – Bolzanella
- VII Gruppo – 26a, 32a, 33a Squadriglia – San Pietro in Gu
- XXII Gruppo – 83a Squadriglia, 2a Sezione SVA – Poianella

Comando Aeronautica VI Armata – Breganze
- II Gruppo - 27a, 48a Squadriglia – Castello di Godego
- VI Gruppo – 76a, 81a Squadriglia – Casoni
- XII Gruppo – 22a, 36a Squadriglia – Casoni
- Gruppo Misto – 21a, 57a Squadriglia – Isola di Carturo

Comando Aeronautica VII e IX Armata – Brescia
- IX Gruppo – 72a, 112a Squadriglia, 6a Sezione SVA – Castenedolo
- XX Gruppo – 74a, 120a, 136a Squadriglia – Ponte S. Marco, 113a Squadriglia – Cividate C.

Comando Aeronautica VIII Armata – Carpenedo
- XV Gruppo – 56a, 115a Squadriglia – Fossalunga
- XIX Gruppo – 23a, 114a, 118a Squadriglia – Istrana
- XXIII Gruppo – 78a, 79a Squadriglia – San Luca

Comando Aeronautica X Armata – Royal Air Force
- 28 Sqn, – Sarcedo, 34 Sqn - Villaverla, 66 Sqn - San Pietro in Gu, 139 Sqn - San Luca

Comando Aeronautica XII Armata – Aeronautique Militaire francese
- SAL 22, SAL 254 – Nove di Bassano

Aviazione Italiana in Francia
- XVIII Gruppo – 3a, 14a Squadriglia Caproni - Longvic

Comando Truppe Italiane in Albania
- VIII Gruppo – 11a Caproni – Tahiraga, 85a Squadriglia – Piskupi
- 116a Squadriglia - Valona

Comando Truppe Italiane in Macedonia
- 73a Squadriglia – Negolani, 111a Squadriglia - Bach

Situation at the End of the War

	Squadrons	Flights	Airplanes	Pilots	Observers	Gunners
Fighter	16		341	320		
Reconnaissance	26		249	297	318	
Strategic recon.	3	5	108	91	19	
Bombers	13		69	167	66	105
Intelligence	1	1	8	14	6	6
SIA 9b	1		11	6	7	2
SVA bombers	2		22	35		
Seaplanes	1		7	5	5	6
UM Squadron	1		5	5	5	
Total	64	6	820	940	426	119

Above: The SVA was an excellent airplane, and it remained in production 8 years after the end of the war. Here are two SVA 5 supplied to the French for evaluation. (Collection S.H.A.A.)

born, the fascist regime had an interest in obscuring the earlier accomplishments, in order to appear as the original creator of Italian air might, so that much of the valor and power of the Army and Navy airmen of the Great War was forgotten.

The Aviation of the Regia Marina, Italy's Royal Navy

Right: The two bastions of Venice: San Nicolò with the airfield, and behind it the Sant'Andrea canal on Le Vignole Island, base of Stazione Miraglia.

The admirals of the Regia Marina, the Royal Navy of Italy, being familiar with the importance of rapid technical innovation at sea, were more open to aviation than the Army generals, who still thought in terms of bayonets and cavalry charges. In 1912 some Italian naval officers got their brevet on seaplanes at the Curtiss flying school in Juan-les-Pins, France. In March 1913 a Sezione Aeronautica (Aeronautical Section) was established at the Office of the Chief of Staff, marking the official birth of Naval Aviation. A Flying School was created at Venice in August 1913 with commander TV Manlio Ginocchio and different seaplanes were procured: Curtiss, Borel, Breguet and Farman Savoia. In 1914 seaplanes were embarked on battleships, experimenting with wireless RT communications. In March 1914 the Sezione Aeronautica became V Reparto (5th Unit) of the Naval Staff, reporting to the General Direction of Naval Construction. However, the interest in developing the new weapon was not supported by an industrial base.

In the summer of 1914 seaplanes were retired from major ships, while an old protected cruiser, RN *Elba*, was converted as a seaplane tender, with commander CF Lodovico De Filippi. In June the first Albatros W.1 arrived, a good German floatplane.

In the following ten months of neutrality, realizing that if war came, it would be against Austria, not against France, important works upgraded the port of Brindisi, the only main base in the Adriatic Sea besides Venice, and Ancona, and work was started to prepare seaplane stations at Porto Corsini, Pesaro and Brindisi. War was already raging in the Adriatic, between the French and Austrian fleets, and Italian seaplanes flew to check for mines.

Funding arrived, and many Curtiss and six Albatros seaplanes (built in Italy) were ordered. In February 1915 a flying school was established at Taranto, with the American John Lansing Callan as chief instructor.

When Italy entered the war, 24 May 1915, there were six seaplanes at Venice, two Albatros, two Borels and two Breguets, with four pilots. Some Borels were based in Porto Corsini, two Curtiss still in assembly were in Pesaro, two more inefficient ones were in Brindisi, the RN *Elba* had three Curtiss and a balloon and the Taranto flying school had 15 Curtiss and fifty students.

The naval convention attached to the secret Pact of London engaging Italy in the war included the deployment of a French naval squadron. Under the command of *Lieutenant de Vaisseau* Antoine Reynaud, and equipped with six FBA type C flying boats, it came immediately in Italy, and flew its first mission on 29 May. A fighter squadron followed in August, Escadrille N 92/I, with commander Capt. Marcel de Chalonge, to oppose Austrian raids against Venice. The French airmen gave excellent proof of themselves throughout the war.

The Navy flew a few reconnaissance flights from Venice, Porto Corsini and RN *Elba* at Brindisi in May and June, but soon only the Albatros at Venice remained operational.

The solution to the Italian technical inferiority came unexpectedly from the capture on 27 May of an Austrian Lohner, L40, which was sent to Nieuport Macchi in Varese who, with an admirable commitment, managed to copy it in a few weeks and start the production of 48 units. In July the Italian-built Albatros with Isotta Fraschini 160 hp engines arrived, and in October the first "L", that is, Lohner Macchi.

Above: Regia Nave *Elba* was the first seaplane tender of the Regia Marina, here entering the harbor of Taranto, with a Curtiss flying boat on its deck.

Efficient Aircraft on 15 August 1917

Station	***Unit***	***L1***	***L2***	***L3***	***FBA***	***MM***	***Sopwith***
Grado	253ª	-	-	10	1	2	-
Venezia	251ª, 252ª, 259ª Sq.	-	-	20	9	1	4
Porto Corsini	12ª Sezione	-	-	-	4	-	-
Varano	254ª Sq.	-	1	8	-	-	-
Brindisi	255ª Sq.	-	-	8	12	-	4
Otranto	25 ª Sq.	-	-	5	-	-	-
S. Maria di Leuca	Sezione	-	4	-	-	-	-
Taranto	Scuola	6	1	-	4	-	-
Milazzo	8ª Sezione	-	-	-	3	-	-
Palermo	3ª Sezione	-	-	-	3	-	-
Napoli	13ª Sezione	-	-	-	2	-	-
Piombino	5ª Sezione	-	-	-	3	-	-
La Spezia	7ª Sezione	-	-	-	6	-	-
Porto Maurizio	9ª Sezione	-	-	-	6	-	-
Sanremo	1ª Sezione	-	-	-	4	-	-
RN Europa	258ª Sq.	-	-	-	7	-	-
Valona	257ª Sq.	-	-	-	14	-	-
Total		6	6	51	78	3	8

This aircraft changed the situation, the Navy now being capable of offensive missions.

In October the entire Navy force, entrusted to 19 pilots, consisted in Venice of a squadron of six Albatros, worn out by long use, and the French detachment. In Grado, an outpost north of Venice, close to the frontline, there were two Borels, in Porto Corsini four Borels and a Breguet, nothing in Brindisi, and four Curtiss on the RN *Elba*. Some French Nieuports were based in Cascina Farello, protecting the lagoon of Venice. In the autumn RN *Elba* was disarmed and the RN *Europa* joined the fleet. This seaplane tender carried eight seaplanes and could operate as an autonomous base. Its seaplanes went into action over the lower Adriatic on 17 October, carrying out daily reconnaissance flights. In

Above: The American John Lansing Callan, instructor for the Italian Navy at Taranto.

War Missions of the Aircraft of the Regia Marina
Number of Missions – in Brackets, Number of Sorties

Year	1915	1916	1917	1918	Total
Bombing	13 (12)	96 (42)	686 (179)	1.382 (329)	2.177 (492)
Reconnaissance and escort	46 (42)	130 (74)	875 (389)	2.416 (891)	3.467 (1.396)
Defense and fighters	14 (9)	50 (21)	261 (103)	782 (298)	1107 (431)
Esplorazione	232 (198)	1.334 (721)	3.704 (1.679)	4.173 (1.778)	9.433 (4.376)
Defense of sea traffic	-	-	2337 (2337)	8048 (8048)	10.385 (10.385)
Totals	305 (261)	1.600 (858)	7.863 (4.687)	16.814 (11.244)	26.569 (17.050)

September the Directorate of Aeronautical Services of the Upper Adriatic was established, under the command of CC Giulio Valli, who was replaced in April 1916 by CF Arturo Ciano.

At the end of 1915 in Venice there were four efficient Macchi L.1s and three Albatros, in Grado, together with the French FBAs, two L.1s under the STV De Rossi, in Porto Corsini, with commander TV Vivaldi Pasqua, four L.1s, but at the beginning of January this base stopped operating, remaining only as a fueling point. In Brindisi, with TV Surdi, there was an L.1 and on the RN *Europa* a squadron under the command of TV Roberti with four L.1s, while a station at Varano, in Apulia, was under construction.

In January 1916, despite the weather, the missions increased: 22 from Venice, plus the same number from the French, about ten from Grado, 22 from Brindisi and 16 from RN *Europa*, still stationed in Brindisi, all reconnaissance, minus four useless attempts to intercept enemy aircraft over Venice

On 31 January 1916 the RN *Europa* moved from Brindisi to the bay of Valona (now Vlora), Albania, with its four L.1s, providing a first example of air-naval power projection and a response to the Austrian seaplanes that from Cattaro (now Kotor) attacked the Albanian port. The seaplanes of the RN *Europa* now had an objective, Durazzo (now Durres), that the Austrians had captured on 23 February, that was bombed on 20 March. The Austrian advance stopped north of Valona and for over two years the Albanian front remained stationary.

In the spring the rate of operations increased, with 95 sorties flown, and the first war flight also took off from the new base at Varano.

The V Reparto of the General Staff was united together with the Submarine Department, forming a new *Ispettorato*

Above: The Austrian Lohner L.40 that fell in Italian hands and was copied by Macchi, here at Porto Corsini; behind it parts of a Borel monoplane can be seen (USMM)

Sommergibili e Aviazione (Inspectorate of Submarines and Aviation), activated in May, under the command of Rear Admiral Mario Casanova Jerserinch until February 1917, then of Rear Admiral Giuseppe De Lorenzi. The head of the aeronautical department was CV Guido Chelotti until January 1917, then CF Lodovico De Filippi.

In June 1916 the *Ispettorato Sommergibili e Aviazione* created two *Direzioni dei Servizi Aeronautici*, (directorate of aviation services) one in Venice, controlling the higher Adriatic down to Jesi, and one in Taranto, controlling the four bases at Varano, Grottaglie, Taranto and Brindisi.

As for training, some Navy pilots began at Army schools, and then moved to Taranto, where towards the end of 1916 the use of American instructors ended, but in 1916 the school, despite the high value of the chief instructor, the civilian Giuseppe Rossi, was not very productive due to the poor quality of the Curtiss seaplanes, and only improved with the arrival of the first Macchi seaplanes. Another school was then established in Sant'Anna di Sesto Calende, under control of the Army, managed as a contractor by the local SIAI company with FBA C seaplanes, which began courses in February 1916, attended first by 140 Navy students, half of whom did not complete the course.

The growth of this effective force, however, was hampered by the Navy's lack of tools to deal effectively with the industry. The existing seaplanes at the aviation stations were 19 on 1 October 1915, 28 on 1 January 1916, 38 on 1 April, 36 on 1 June, 45 on 1 September, too low a number after 15 months of war. The problem of the lack of pilots, especially officers, also remained.

The solution was a Decree of 7 September 1916 which placed the naval air service under the Ministry of War. As a consequence, responsibility for aeronautical construction passed to the Ministry of War, which took over the technical part of aeronautical services. From this moment, the new planes no longer had Navy serial numbers, but unique ones with those of the Army, a system that was maintained throughout the war.

This choice was due not only to overcome industry delays but also to the Navy's difficulties in detaching officers from ships to the flight services, to which Army pilots would now also be assigned. The Navy was left with just the task of providing the observers and obviously determining the use of the air assets. Austria followed the opposite path, establishing in August the *K.u.k. Seeflugleitung* (Naval Aviation Command).

This innovation had negative effects on the morale of the Navy officers, who shortly afterwards saw Army colleagues arrive and even being assigned command of one of the two squadrons formed in Venice in December. The coexistence between the two services turned out to be positive; the Army pilots lacked neither ability nor commitment, but war at sea required a body of specific knowledge that only seafarers had from their own personal professional experience.

Above: The unlucky Bresciani bomber seaplane at Sant'Andrea.

Personnel at the Naval Air Stations

Date	May 1915	December 1916	December 1917	November 1918
Personnel of the Regia Marina				
Pilots	11	62	119	227
Assigned manforce	128	747	1,839	3,098
Personnel of the Regio Esercito				
Pilots	-	1	79	278
Assigned manforce	-	3	1,274	3,132
Totals	139	813	3,232	6,457

At the beginning of 1917 the Aviazione della Marina had the *Direzione dei Servizi Aeronautici* of Venice (commander first the CF Arturo Ciano, then CC Emanuele Ponzio) with a squadron in Grado, with FBA and then L.3, two in Venice, 1st with FBA and 2nd with Lohner, one, non-operational, in Porto Corsini, and the *Direzione dei Servizi Aeronautici* of Taranto, commander CF Salvatore Denti di Pirajno, with a squadron in Brindisi, one in Varano, both with FBA, one on the RN *Europa* in Valona and one in Taranto.

In February 1917, Macchi L.3s and FBAs were procured for a new station in Otranto and new flight schools were established in Orbetello and Passignano, in central Italy, both equipped with L.1s, managed by the Army.

One of the first acts of Admiral Paolo Thaon di Revel, who on 4 February 1917 returned as Chief of Staff of the Navy and Commander in Chief of the Naval Forces, was to review the rash decision of 1916 and bring naval aviation back under the direct control of the Navy. The control of the Ministry of War was maintained, through the Technical Directorate of the Air Force, over production and supplies, appropriately unifying controls and procedures, and restoring the Navy's control missions at the manufacturing companies. The Taranto flight school, which had stopped with the transition to the Army, was reactivated.

In March 1917 a new squadron, with Navy personnel was activated in Otranto, initially with six L.3s and another one, with Army personnel in Valona, initially equipped with 15 FBAs (demonstrating that production coordination had worked, and there was now no shortage of machines) next to the one embarked on the RN *Europa*, with Macchi Lohners. In June the Porto Corsini squadron was reactivated, with FBAs and a section of L.3s was activated in Santa Maria di Leuca, the the most southern point of the Adriatic. The arrival of new seaplanes allowed the transfer of old aircraft to the schools, so another school was established in Varese, with some L.1s, while the one at Passignano received ten L.1s, and even three archaic Curtiss Colombos.

In the spring of 1917, the *Ispettorato Difesa del Traffico*

Above: The FBA, license built from France, was the backbone of Italian naval aviation, over 1,000 were produced by six factories, and it was even put back in production after the war. This one, 3181, is at Brindisi in the first half of 1916.

Above: Regia Nave *Europa* could accommodate eight seaplanes; it operated very effectively from the harbor of Valona, Albania.

Aircraft of the Italian Navy in 1918

Stazione	January 1918	June 1918	Existing in October 1918
Venice	24 L.3, 25 M.5, 2 ISVA	36 L.3, 25 M.5, 3 ISVA, 4 HD	34 L.3, 24 M.5, 9 M.8, 1 M.9
Porto Corsini	15 FBA	14 FBA, 8 M.5	6 FBA, 6 M.5, 12 M.8
Ancona	13 FBA	3 FBA, 4 M.5	13 FBA, 3 M.5, 2 M.8
Varano	8 L.3, 1 L.2, 6 FBA, 4 M.5	12 FBA, 2 L.3, 4 M.5, 1 Sopwith	12 FBA, 8 L.3, 2 M.8, 28 M.5
Brindisi	13 FBA, 3 L.3, 8 M.5, 1 Sopwith	13 FBA, 2 L.3, 5 M.5, 1 Sopwith	12 FBA, 8 L.3, 2 M.8, 13 M.5
	4 Nieuport, 5 SP.3, 2 SP.2	5 Nieuport, 4 SP.3	4 Nieuport, 2 SP.3, 2 SVA
Otranto	11 FBA	5 FBA 4, M.5	13 FBA, 11 M.5
S. M. di Leuca	4 FBA	-	–
Taranto school	7 L.1, 4 L.2, 4 L.3, 1 FBA	14 FBA, 3 L.1, 1 L.2, 4 L.3, 6 Sop.	24 FBA, 10 L.3, 3 ISVA
Valona	15 FBA, 2 M.5	12 FBA, 3 M.5, 4 ISVA	13 FBA, 4 M.5, 3 ISVA
RN Europa	8 FBA, 2 M.5	11 FBA, 5 M.5	13 FBA, 4 M.5, 3 ISVA
Taormina	3 FBA	4 FBA	9 FBA
Catania	4 FBA	4 FBA	5 FBA
Syracuse	3 FBA	4 FBA	4 FBA
Trapani	2 FBA	6 FBA	8 FBA
Palermo	4 FBA	13 FBA	10 FBA
Milazzo	5 FBA	5 FBA	4 FBA
Sapri	4 FBA	6 FBA	4 FBA, 3 S.8
Naples	2 FBA	3 FBA 2 M.5	9 FBA, 2 S.8
Ponza	3 FBA	5 FBA	5 FBA
Civitavecchia	4 FBA	6 FBA	2 FBA, 3 S.8
Piombino	4 FBA	5 FBA	5 FBA
Campiglia	2 Caproni 350	2 Caproni 350	-
Livorno	5 FBA	3 FBA	6 FBA
Spezia	5 FBA	8 FBA	7 FBA, 1 S.8
Rapallo	-	9 FBA	10 FBA, 4 Sopwith
Varazze	2 FBA	-	-
Porto Maurizio	5 FBA	7 FBA	8 FBA
Sanremo	4 FBA	10 FBA	3 FBA, 7 S.8
Cagliari	4 FBA	-	3 FBA
Terranova P.	2 FBA	3 FBA	5 FBA
Bolsena school	-	3 FBA, 2 L.3	55 FBA, 7 L.3, 1 M.8, 11 M.5, 2 ISVA
Sarzana	-	3 Nieuport 80 hp	4 SVA
Civitavecchia	-	8 FBA, 1 ISVA	4 S.8, 1 ISVA
Ferrara	-	6 Nieuport 80 hp	5 Nieuport 110 hp, 5 SVA
Jesi	-	2 SP.2, 1 SP.3, 1 Farman	k4 SP.3, 4 SVA, 2 Farman
241ª Squadriglia	-	-	13 Hanriot, 1 Nieuport
Orbetello	-	-	5 FBA, 5 S.8
Poggio Renatico	-	-	11 Ca.4, 14 Ca.5
1ª S.A.	-	-	3 SIA 9B, 2 Ca.5, 2 SVA
San Vito	-	-	1 Ca.5
Totals	259	346	593

Marittimo Nazionale (Inspectorate for the Defense of National Maritime Traffic) was created at the Ministry of the Navy, an agency that effectively coordinated all forces and equipment in the anti-submarine war.

In the spring of 1917 the naval aviation units acquired the official name of "squadriglia," with a number, like the Army units. Seaplane units were assigned numbers from 251, while land-based aircraft units would be numbered from 201. The 1st L.3 Squadron of Venice, became the 251ª Squadriglia, the 2nd FBA Squadron, 252ª, the FBA and L.3 squadron of Grado 253ª, the unit at Varano, with Macchi Lohner, 254ª. In Brindisi the FBA unit became the 255ª Squadriglia, the unit in training in Otranto with L.3s, became the 256ª Squadriglia on 1 July, the FBA unit in training for Valona became 257ª Squadron in May, while the L.3s of the RN *Europa* formed the 258ª Squadriglia, which in those days was being re-equipped with FBAs.

The attacks on Venice and Trieste led to an exchange of mutual accusations between Italians and Austrians, taken up by the international press, and on 11 June the Ufficio Servizi

Left: CF Salvatore Denti di Pirajno and Giulio Valli, commanders of aviation services in the southern and northern Adriatic.

Aeronautici of the Supreme Command wrote (signed by the deputy Chief of Staff of the Army, General Porro) to Revel requesting that Trieste no longer be bombed.

In June 1917 the heroic French unit, the *Centre d'Aviation Maritime de Venise*, reduced to a minimum after the loss of all its officers, returned home. However, by now the Italian naval air units at Venice were fully capable of supporting the conflict on their own. A third Venice squadron, the 259[a], was then formed on 22 June with six L.3s.

In August 1917 the V Reparto, still under the command of CF De Filippi, approved a new development program, which envisaged an order of battle with around fifty offensive squadrons, for a total of 400 seaplanes and 32 bombers and 56 land fighters. To these were added another 150 seaplanes for the defense of traffic on the Tyrrhenian Sea. This program, like all the ones drawn up by all air forces during the Great War, was too ambitious in terms of producing aircraft and even more of training crews.

The crucial point in the Mediterranean was the Strait of Otranto, between Italy and Albania, which Austrian and German submarines had to pass through from Cattaro and other bases for missions in the Mediterranean. A barrier was established, first with nets lowered by English fishing boats called drifters, then with fixed nets. In June 1917 a common system of patrolling the lower Adriatic was established with the Allies, with French seaplanes in Corfu, English seaplanes and Sopwith fighters in Otranto, and, as an Italian contribution, 26 seaplanes in Brindisi, five in S. Maria di Leuca, nine in Otranto, 22 FBAs in Valona, 18 seaplanes in Varano, plus some airships, each group with its assigned patrol area.

At the end of July 1917 there were 113 efficient aircraft: in Grado eight L.3s and two FBAs, in Venice 13 L.3s,

Below: The Macchi L.3 flew better but floated worse than the FBA, and was built in smaller numbers, just 203. This is 3425 in Venice.

Above: Pola today, now Pula in Croatia, was the constant target of Italian bombings. The seaplane station was in Santa Caterina island, in the center of the bay.

12 FBAs and one M.M., in Porto Corsini three FBAs, in Varano one L.1 and six L.3s, in Brindisi seven L.3s and eleven FBAs, in Otranto four L.3s, in S. Maria di Leuca three L.3s. Across the Adriatic, there were eight FBAs on the RN *Europa* and nine FBAs in Valona. The Taranto school had one L.1, three L.2, and three FBAs.

In the summer of 1917, a small unit of Army land fighters was formed, the Sezione Difesa Nieuport in Grottaglie, an airship base close to Taranto, while another Defense Section was in Brindisi, both at the end of October 1917 directly assigned to the Navy.

The program presented by the Inspectorate in 1917 aimed to create three groups of Caproni land bombers by the spring of 1918, each with 80 aircraft, an absolutely unrealistic program, considering that the Army deployed just 50 Capronis. Meanwhile, all the Regia Marina could get was the 201ª Squadriglia with four Caproni Ca.3s with Army personnel, based at Marcon, near Venice.

In 1917, the seaplanes at Venice made 3,440 flights, shot down four enemy aircraft while 25 seaplanes were lost or destroyed. At the end of the year there were 34 aircraft in Venice, 7 in Porto Corsini, 4 in Ancona, 11 at Varano, 22 at Brindisi, 6 at Otranto,4 at Santa Maria di Leuca, 24 at Valona and 8 on RN *Europa* plus 20 at the school at Taranto.

The unit at Brindisi, the 255ª Squadriglia, now too large, was split into three units, giving birth to the 262ª Squadriglia, with M.5 fighters, and the 265ª Squadriglia, with Macchi L.3s. With considerable flexibility, the operational squadrons in the south, like the ones in Venice, also acted as flight schools, eventually sending the students to Taranto for formal tests to obtain the 2nd license, that is, the brevet of military pilot. The three squadrons at Venice specialized in bombing, 251ª, 253ª and 259ª, formed a Group, with as commander TV Enrico Accorretti.

A new flight school was created in Bolsena to train US Navy students, under the command of TV Mario Calderara. Initially equipped with FBAs, and with Navy instructor pilots, both Italian and American, it began flights on 23 February 1918 with around sixty American students, who were joined in the spring by the same number of Italians and in April it issued the first brevet. In June, when it was equipped with 48 FBAs, the school also received M.5s to provide fighter training and in July the first American pilots were able to leave for the front and their places were taken by other Italian students. Overall, the Bolsena school carried out 11,597 flights with five fatal accidents and awarded 102 first brevets and 104 second brevets.

After the death of CC Emanuele Ponzio due to a car accident on 6 April, the *Direzione Servizi Aeronautici* of the upper Adriatic was assigned again to CF Giulio Valli,

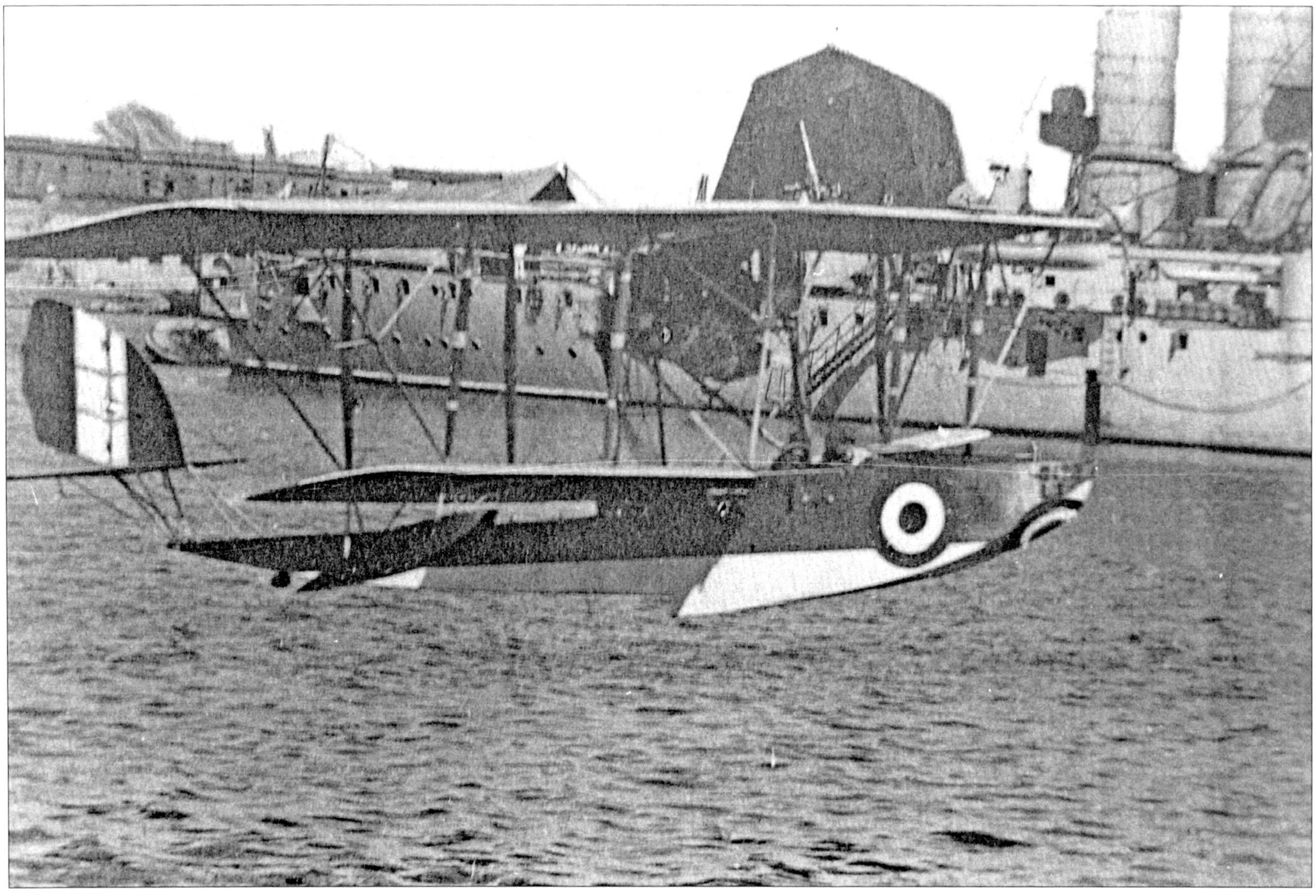

Above: A Macchi L.3 taking off from Brindisi next to a battleship.

director of the Aeronautical Services of the lower Adriatic became CV Alberto Viscardi.

A Navy flight school for land-based aircraft was established in Capua Nord, with 300 students, under the command of TV Bruno Brivonesi: the problem of staff shortages was now more serious than that of the shortage of aircraft. It was decided to assign the American pilots trained in Bolsena to the Porto Corsini station and to the stations under construction, in Pesaro and Pescara, and some of the Americans trained in Foggia for the future squadrons of Capronis.

A major action was launched against Pola on 17 July. It began with a night attack by the airships and was continued during the day by twenty Capronis and nine SVAs of the Army, together with 27 seaplane bombers escorted by 21 Macchi M.5 fighters from Venice and finally four SIA 9Bs, with the support of torpedo boats, and of the American FBA squadron from Porto Corsini. It was a level of cooperation between Navy and Army that was not matched in World War 2.

In September the Air Force Command made its XXII Gruppo with the 181ª and 182ª Squadriglia of Ca.4 triplanes available to the Navy. This was not much of a gift, given that the triplanes had been inactive for two months in Ghedi after two fatal accidents. The Ca.4s moved to Poggio Renatico, where work was in full swing on the new base.

The superiority of the Allies was now overwhelming: the Regia Marina alone now had four times as many aircraft in service and, above all, an almost ten times as many pilots than the Austrians. It is estimated that in September 1918 the *K.u.k. Seeflieger* had 134 planes in line, 27 in Trieste with 12 pilots, 53 in Pola with 18 pilots, six in Sebenico (Šibenik) with six pilots, 45 in Kumbor with 12 pilots.

On 24 September 1918 a decree established the *Ispettorato dell'Aeronautica*, Inspectorate of the Air Force, taking on the functions that belonged to the *Ispettorato dei Sommergibili e dell'Aviazione* and having control also of the Aerostatic Service. CV Giuseppe Marsilia took the place of CF De Filippi, then Rear Admiral Pietro Orsini.

In 1918 the Miraglia Station in Venice, with a force of around sixty seaplanes and the same number of pilots, had carried out 4,592 flights in 1,080 war missions, including 37 attacks on Pola with formations of up to 25 aircraft. During the war, aircraft and airships attacked Pola 55 times, from 30 May 1915 to 22 October 1918, dropping over 83 tons of bombs.

Overall, the naval aviators were awarded two Gold Medals for Military Valor (Garrassini Garbarino and Casagrande), 165 Silver and 172 Bronze Medals. According to the official report, among the aviators there were 102 killed while 37 planes were lost due to enemy action, 77 due to failures during war missions and 290 due to accidents.

Above: One of the first Macchi MM of the first production batch of ten, 4873, at Varese, in the cockpit is Macchi test pilot Giacomo Reggiori who died in a test flight on this aircraft on 19 July 1917.

Above: Admiral Paolo Thaon di Revel, commander of the Italian Navy during the First World War.

Above: CC Emanuele Ponzio, chief of aviation services in Venice, center, in a dark uniform, and Gabriele d'Annunzio, with a cape, meeting French officers at San Nicolò, Lido di Venezia.

Above: A large model of the bay of Pola was built in Venice; here CF Valli in the center, and TV Roberto, right, are showing it to a visiting Admiral.

Right: A postcard showing the emblems of the squadrons of the Stazione Miraglia, Venice.

Facing Page, Below: US Navy officers at Bolsena, celebrating the establishment of the flight school. In the center Cdr. Charles R. Train, Naval Attaché in Rome and commander of US Naval forces in Italy, with Mario Calderara. The officer standing, second from the left, is Lieut. Ernest J. King, later Commander in Chief, US Naval Fleet in World War 2.

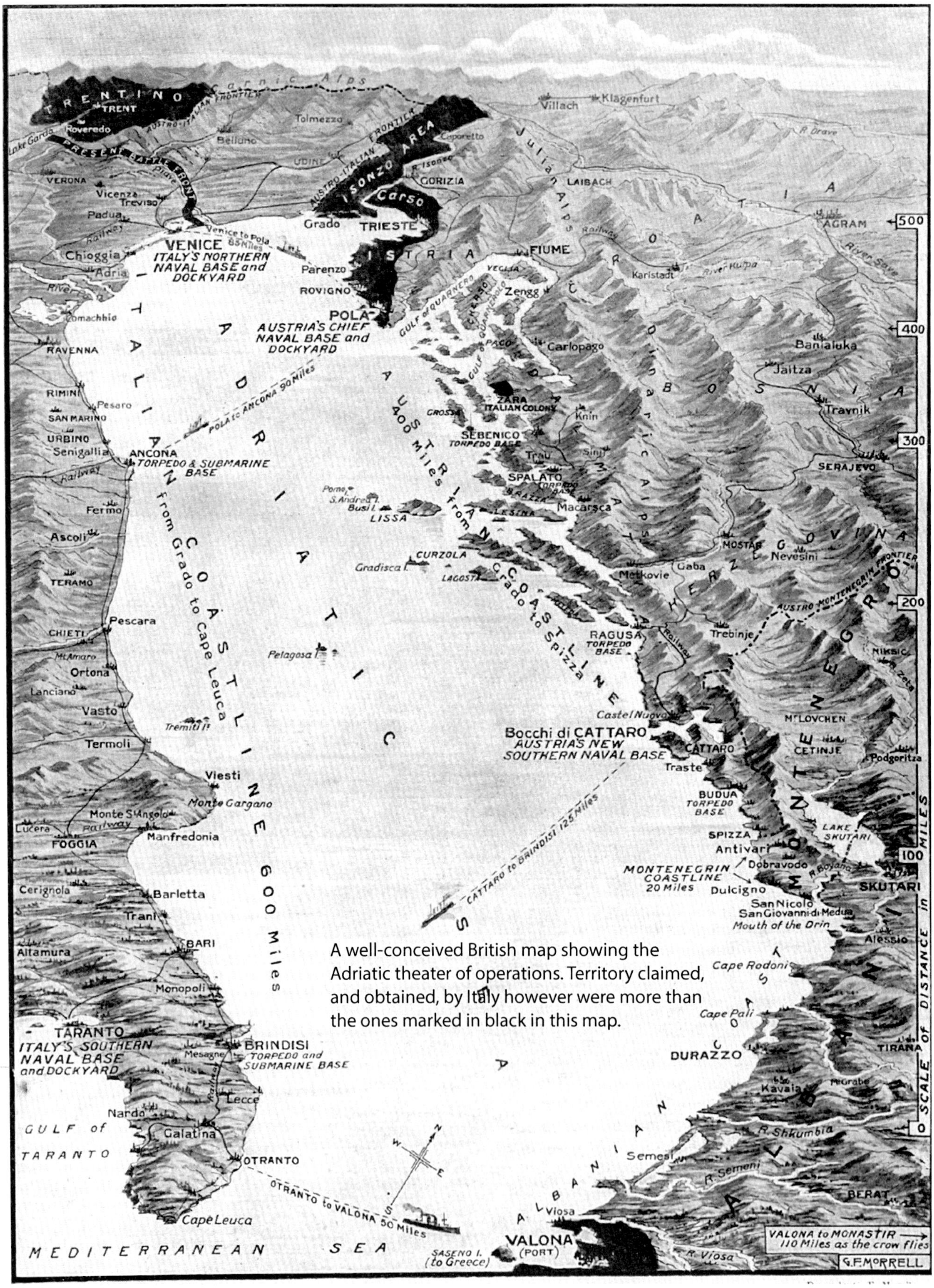

A well-conceived British map showing the Adriatic theater of operations. Territory claimed, and obtained, by Italy however were more than the ones marked in black in this map.

Italian Aircraft Profiles courtesy Marco Gueli

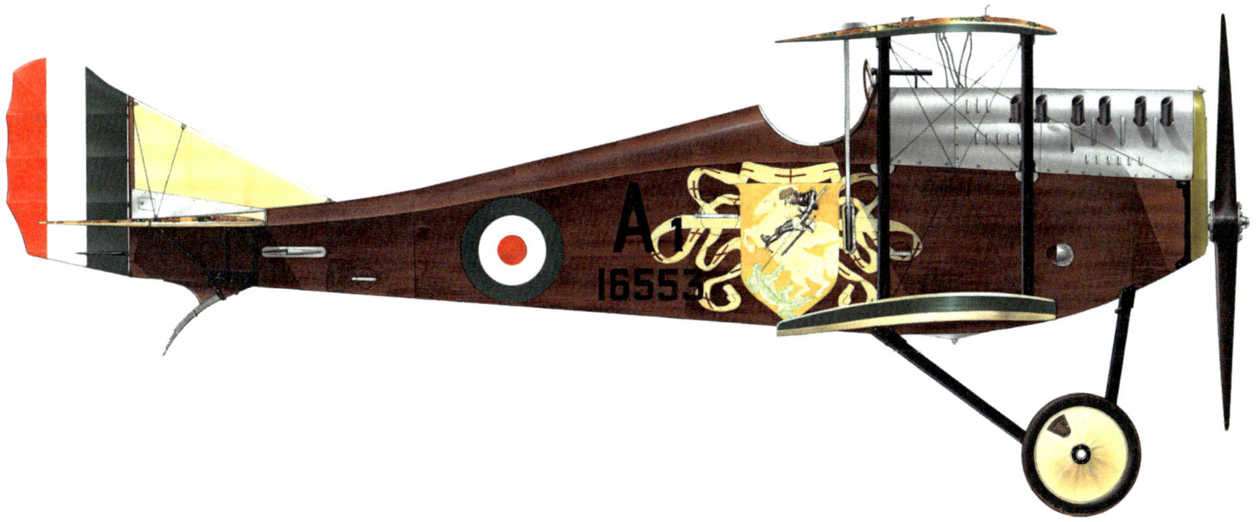

Ansaldo A.1 Balilla
Airplane gifted to *ten*. Antonio Locatelli
Genoa, August 1918

Caproni Ca.1 #1151
8ª Squadriglia, crew Pagliano, Gori, Barbarisi
La Comina, 1917

FBA Type C 61
1ª Squadriglia Idrovolanti
Desenzano del Garda, Autumn 1916

Hanriot Hd.1 #7501
85ª Squadriglia
Albania, 1918

Hanriot Hd.1 #515
***Ten*. Baracchini, 76ª Squadriglia**
Casoni 1918

Hanriot Hd.1 #15309
Gifted to *ten*. Baracchini
1919

Nieuport 10
1918

Spad 7 #14208
77ª Squadriglia, brig. Cabruna
Marcon, 1918

Spad 7 #153
91ª Squadriglia, *ten*. Ruffo di Calabria
Santa Caterina, 1917

Caproni Ca.3 #11505
1ª Squadriglia, *magg*. Bonazzi
San Pelagio, 1918

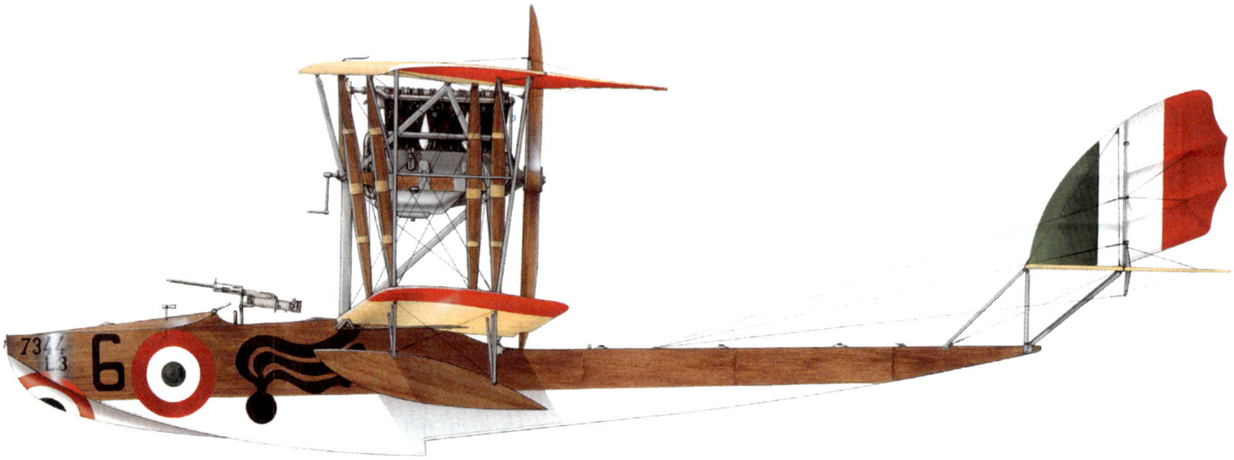

Macchi L.3 #7344
251ª Squadriglia
Venice, 1917

Macchi M.5 #7262
261ª Squadriglia, STV De Riseis
Venice, 1918

Nieuport 17 #2614
91ª Squadriglia, *cap*. Baracca
Santa Caterina, 1917

Nieuport 27 #19750
79ª Squadriglia, *serg*. Cerutti
San Luca, October 1918

Pomilio PE #7020
112ª Squadriglia
Castenedolo, 1918

Spad 7 #1673
77ª Squadriglia, *serg*. Ferrario
Marcon, 1918

SAML S.2 S.1904
113ª Squadriglia
Cividate Camuno, 1918

SIA 7B #5803
36ª Squadriglia
Istrana, December 1917

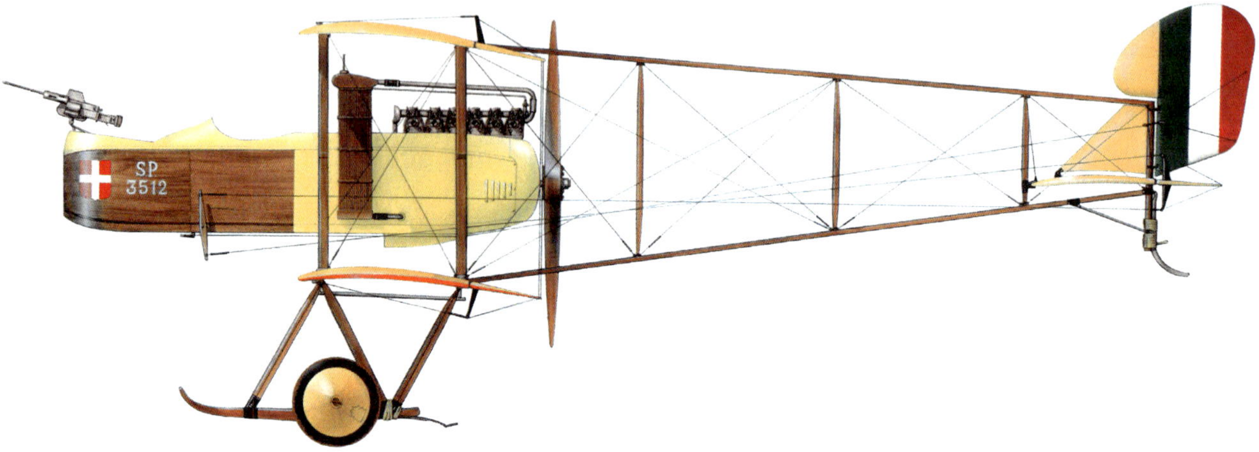

SP.2 #3512
21ª Squadriglia
Campoformido, 1917

Ansaldo SVA #12736
87ª Squadriglia, Palli and d'Annunzio
San Pelagio, August 1918

Part 3: The Gruppi

Above: Nieuport 11s of the 80ª Squadriglia and SP.2 of the 23ª Squadriglia, I Gruppo. This line-up of aircraft is not for operations, but for some sort of ceremony.

The Gruppi were medium level units, providing the command for several squadrons. They could be formed by squadrons of different specialties, or with a single mission, like bomber groups. They formed the administrative link between the Aeronautical Commands of each Army and the Supreme Command, and the squadrons. Group commanders, particularly the ones promoted from squadron commands, were often very popular figures, some of them flying a great number of combat missions.

I Gruppo

The I Gruppo was formed on 28 March 1913 with Squadriglie 1ª and 3ª Blériot, 2ª Bristol. At the outbreak of war, on 24 May 1915 it deployed to Portogruaro under the command of *cap.* Alberto Novellis di Coarazze. At that time, it consisted of Squadriglie Blériot 1ª, 2ª, 3ª, 13ª and 14ª. It was assigned to the 3ª Armata, southern part of the front, with its 3ª Squadriglia at San Vito al Tagliamento, 13ª and 14ª at Torresella (Giussago). In June it moved to Mortegliano and then to Aiello, losing its 1ª and 14ª Squadriglia, disbanded on 22 June. During the First Battle of the Isonzo it fought having its 2ª at Chiasiellis, 3ª at Quota 43 and 13ª at Pozzuolo. It gave the 2ª to the II Gruppo. In July it moved to Cervignano, receiving the 11ª Farman. These are all villages in the southern part of the plains of Friuli. On 1 August 1915 also its 3ª and 13ª were disbanded so the group included only the 11ª, and for only 12 days the 2ª was also attached.

On 17 August 1915, after further movements, it was at Cervignano, comprising the 11ª at Chiasiellis out of action and the new 5ª Voisin at Campoformido. Then it moved to Mortegliano and on 28 August it got the 7ª Voisin, being thus now formed by the Squadriglie 11ª Farman, 5ª and 7ª Voisin, all at Chiasiellis. On 13 September the Gruppo incorporated the 10ª Farman of S. Maria la Longa from the disbanded III Gruppo while *magg.* Augusto Gallina became commander. Now based at S. Maria La Longa, formed by 5ª and 7ª Voisin, 10ª and 11ª Farman, the Gruppo kept this strength until January 1916, when it received the 2ª Squadriglia Caccia at Cascina Farello, Aquileia. In March *magg.* Alberto Marenco became commander.

On 15 April 1916 its Squadriglie 5ª, 7ª, 10ª and 11ª Ricognizione and 2ª Caccia were re-numbered as 25ª, 26ª, 27ª, 28ª and 71ª. In May the I Gruppo received the new 36ª, followed on 29 May by the 76ª that remained until 29 September. On 8 July 1916 the 71ª went to the III Gruppo, and 77ª Nieuport replaced it, while *magg.* Oronzo Andriani became commander. On 15 August the 27ª went to the II Gruppo, and for the rest of the year the I Gruppo had 25ª, 26ª, 28ª, 36ª and 77ª. In January 1917 it received the 2ª Idro based at Grado. On 24 February the 26ª left the front to re-equip.

In March 1917 the I Gruppo received three new squadrons: 38ª SP at Risano, 80ª Nieuport at S. Maria la Longa and 39ª SP at Sammardenchia. Thus, in April, always

Above: Bleriots and a Farman of the III Gruppo at an airfield in the flat countryside south of Udine.

assigned to the 3ª Armata, its structure consisted of 25ª Voisin at Pozzuolo, 36ª SP at S. Maria la Longa, 38ª SP at Risano, 39ª SP at Sammardenchia, 77ª and 80ª Nieuport at Aiello, 2ª Idro at Grado. In April 1917 it received the new 23ª SP at S. Maria la Longa. On 5 May 1917 it gave its 38ª and 39ª Squadriglia to the V Gruppo, receiving a Spad Flight based at S. Maria la Longa.

On 3 July 1917 the I Gruppo, still at S. Maria la Longa, received the new 84ª Squadriglia Caccia and the 112ª SAML at Lavariano, coming from Villaverla. It got on 25 July also the new 131ª Pomilio. Command was then assigned to *t.col.* Novellis di Coarazze, who had been its commander in 1915. In autumn, I Gruppo had Squadriglie 23ª, 25ª, 77ª, 80ª, 84ª, 112ª, 131ª and 2ª Idro. On 15 October the 112ª was assigned to the V Gruppo, and *cap.* Luigi Sella became temporary group commander.

During the Caporetto retreat the I Gruppo was left with just 25ª at Arcade and 77ª and 80ª at La Comina. In November *cap.* Arnaldo Bevagna was the commander, until 20 December, when *cap.* Sella became full-time commander. The unit received the 117ª SAML that moved from S. Giustina to Padova, so that now it included 112ª SAML, 117ª SAML and 131ª Pomilio.

On 4 February 1918 the I Gruppo moved to Fossalunga. In March the 112ª was made available for operations to the V Gruppo, and on 19 March this squadron went to the IX Gruppo. On 14 May the I Gruppo received the new 5ª Sezione SVA, but at the end of the month its 131ª went to the V Gruppo. The I Gruppo, always assigned to the 3ª Armata now included only the 117ª and 5ª Sezione SVA until 7 July, when the 131ª returned and the 117ª was disbanded.

On 4 November 1918 the I Gruppo was at Carpenedo, Venice, with the 131ª and the 5ª Sezione SVA. On 21 November it got the 77ª, 131ª and 3ª Sezione SVA Biposti. In December 1918, still with *cap.* Sella as commander, the unit was at Aquileia with 77ª, 131ª, 3ª Sezione SVA, all at Cascina Farello. This structure remained in 1919, while *cap.* Mario De Bernardi became its commander. On 28 February 1919 the Gruppo moved to Zaule, Trieste, losing the disbanded 131ª. In May 1919 also the 77ª Squadriglia and the I Gruppo itself were disbanded.

II Gruppo

At the outbreak of war, the II Gruppo consisted of Squadriglie 5ª, 6ª and 8ª Nieuport. It mobilized at Campoformido with 6ª, 7ª and 8ª Nieuport at the command of *cap.* Lelio Gaviglio, assigned to the 2ª Armata. It received from the I Gruppo the 2ª Squadriglia Blériot, that was soon disbanded, then it gave its 6ª to the III Gruppo and received the 9ª Farman, so until July it operated with 7ª and 8ª Nieuport and 9ª Farman.

On 13 August 1915 the II Gruppo got the 5ª Voisin,

Above: Caproni Ca.3 of the 3ª and 7ª Squadriglia of the IV Gruppo detached from La Comina – Pordenone to Aviano for a ceremony in July 1917.

passing it four days later to the I Gruppo. The 7ª Nieuport was disbanded on 17 August, as the 2ª Farman replaced it, but on 21 August this unit was assigned to the Supreme Command. Overall, from 24 May to 31 July 1915 the three units of II Gruppo, Squadriglie 7ª, 8ª and 9ª, flew 117 reconnaissance flights lasting 178 hours, dropped 104 bombs and 4,400 darts and took 167 photographs.

After the disbandment of the Nieuport units, the II had just the 9ª Farman. On 25 September it received the new 6ª Farman that moved to Chiasiellis on 6 October. On 31 January 1916 it got the new 4ª Farman. After the death of *t.col.* Barbieri on 18 February commander Gaviglio went to the Gruppo Comando Supremo while *cap.* Ercole Capuzzo took command of the II Gruppo. Its units covered the "Gorizia zone" of the front of the 2ª Armata.

In March the 9ª Squadriglia went to the Gruppo per l'Artiglieria. In April the 4ª became 29ª Squadriglia, the 6ª became 30ª, then II Gruppo got the 35ª Voisin and 76ª Caccia formed on 24 May 1916 but just five days later, on the 29th, the 76ª went away, and also the 30ª went to the III Gruppo, from April to 21 July.

On 15 August the II Gruppo got the 27ª, thus it now consisted of 27ª, 29ª, 30ª and 35ª Squadriglie. On 1 October it took over again the 76ª Nieuport. From the start of the war to the end of 1916 the group's units flew 811 reconnaissance missions, 654 fighter flights and 142 bombings for a total of 3,366 flight hours, losing 21 aircraft for various reasons and claiming four victories (76ª Squadriglia) and its airmen received 17 *Medaglie d'Argento* and 22 Bronzes.

On 2 March 1917 the II Gruppo received the 24ª Squadriglia SP, and ten days later the 40ª and 22ª so it became very large, controlling 22ª, 24ª, 27ª, 29ª, 30ª, 35ª, 40ª and 76ª Squadriglie. The 35ª Voisin was disbanded on 11 March. On 10 April 24ª and 29ª moved from the II to the VI Gruppo, the other Gruppo assigned to the Aeronautics Command of the 2ª Armata.[1] The II Gruppo received the 41ª Squadriglia, and then on 21 April 1917 the 81ª, thus the new structure of the II Gruppo was 22ª, 27ª, 30ª, 40ª, 41ª, 76ª and 81ª. In April the II Gruppo also received the new 21ª and for a few days the 2ª Sezione of 113ª SAML, that in May went to the VI Gruppo, and so did the 41ª.

In May 1917 the II Gruppo was heavily engaged in the 10th Battle of the Isonzo: on 23 May it launched a mass bombing of 26 aircraft escorted by 20 fighters. During the Battle the group flew 1,010 war missions for 1,580 hours. In May its fighter units, 76ª and 81ª formed the Sottogruppo (sub-group) di Borgnano.

In June 1917 the II Gruppo, now formed by 21ª, 22ª, 27ª, 30ª and 40ª Squadriglia plus the two fighter units, 76ª and 81ª, was less active, due to bad weather and to the coming abolition of some units. The 27ª Farman was disbanded on 2 July and a new 27ª with Savoia Pomilios was formed on 21 July. On 15 July the Gruppo received the 132ª Pomilio, on 17 July it lost 30ª Farman, disbanded. On 25 July the new 33ª and 35ª SP arrived, while 76ª and 81ª went to the VI Gruppo. Within the reorganization of groups, the II was now only a reconnaissance unit. In August, still based at Udine and assigned to the 2ª Armata, it had on strength

Above: Italian and American officers in front of the command of the IV Gruppo in late 1918. The second man from the left, sitting, is Federico Bonazzi; the fifth is Ernesto La Polla.

21ª, 22ª, 27ª, 33ª, 35ª, 40ª and 132ª. On 10 August it gave its 35ª to the VI Gruppo. On 2 September it gave its 132ª to the X Gruppo, replaced by the 133ª. On 14 September it received the 114ª from the X Gruppo and gave the 33ª to the VII Gruppo. Its new structure then was 21ª, 22ª, 27ª, 40ª, 114ª and 133ª, and its new commander *magg.* Gino Zanuso. In October the 27ª went to the XII Gruppo, while the II Gruppo received the new 118ª

Like all units on the Isonzo front, the group sent regular reconnaissance flights up to October 23 that failed to spot the coming Austro-German offensive. On 25 October the 132ª returned to the II Gruppo, while 118ª went to V Gruppo, and on 3 November, the 22ª went to XII Gruppo. On 10 November the II Gruppo, commanded by *t.col.* Zanuso, was assigned to the new Comando di Aeronautica of the 4ª Armata, with its 113ª, 114ª, 132ª and 133ª Squadriglia, all based at Casoni, and this structure remained into 1918. These squadrons were equipped with SAML and Pomilio, tasked with photographic and strategic reconnaissance and light bombing.

On 27 February 1918 the 3ª Sezione SVA based at Isola di Carturo arrived. The 113ª SAML went to the IX Gruppo, while the 114ª SAML passed on 31 March 1918 to the new XIX Gruppo, thus in the spring the II Gruppo was formed by 132ª and 133ª Pomilio and 3ª Sezione SVA, all at Isola di Carturo. The command of the group was first at Trucca and then, since May, at Marsango.

On 30 June the 3ª Sezione SVA left the II Gruppo, becoming unattached, while the II Gruppo was totally renewed, receiving the 27ª and 48ª SIA coming from the XII Gruppo, as 132ª and 133ª Pomilio were sent back to the Centri Formazione Squadriglie to be disbanded. Command of the Gruppo moved to Castello di Godego, and on 11 July *cap.* Vittorio Giovine replaced Zanuso as commander. The II Gruppo remained with just 27ª and 48ª until the Armistice, 4 November 1918.

On 1 December both 27ª and 48ª were disbanded, and the group reformed with 21ª SAML and 57ª SVA. From December 1918 it was at Aviano, at the command of *magg.* Djalma Juretigh, formed by 21ª, 80ª and 4ª Sezione SVA. The II Gruppo was disbanded on 13 February 1919.

III Gruppo

Formed in March 1915, the III Gruppo had Squadriglie 7ª, 9ª, 10ª, 11ª, 12ª and 15ª, mostly Farman units. With commander *cap.* Augusto Gallina, it mobilized with 5ª Nieuport, coming from the II Gruppo, at Campoformido, 9ª and 10ª Farman at Pordenone, assigned to the Comando Supremo, and also 6ª Nieuport was ordered to report to this

Above: Cap. Alberto Livi, commander of the VIII Gruppo, with a Sopwith Camel, the type of airplane in which he lost his life.

group for operations. On 2 June the Gruppo was assigned to the 3ª Armata, so it moved its command to Udine, then later in that month it went to S. Maria la Longa receiving the 6ª Nieuport. On 31 July it lost 5ª Nieuport, that was disbanded. In August it had 6ª Nieuport, 10ª Farman and the first Sezione Caproni, but 6ª was disbanded on 24 August and the III Gruppo was left with just 10ª Farman. It then received the new 1ª Farman, that was soon re-assigned for the defense of Brescia, and on 13 September 1915 the III Gruppo was disbanded.

A new III Gruppo was formed on 1 January 1916 at Tombetta, Verona, for the 1ª Armata, operating in the north-western part of the front, consisting of 5ª Caproni, 1ª Farman and 4ª Aviatik at Verona, 12ª Farman at Thiene, 3ª Aviatik at Brescia, and the Sezione Idro at Desenzano. Its commander was *cap.* Ernesto La Polla. On 16 January 1916 this unit had 34 aircraft and 33 pilots, 16 of them officers. Its squadrons since 15 April 1916 got the new numbers 5ª Caproni, 31ª, 32ª, 72ª, 73ª and 1ª Idro. From 24 April to 21 July 1916, it had on strength also the 30ª Farman of the II Gruppo that moved temporarily to Verona. On 25 May it got the new 10ª Caproni but on 4 June this unit went to the IV Gruppo. Then the III Gruppo got 9ª Caproni and 75ª Nieuport at Verona.

The III Gruppo was much engaged during the Austrian offensive of May 1916 and the following counteroffensive. On 8 July the 71ª Nieuport at Villaverla joined it, thus from the summer 1916 to spring 1917 it was formed by 5ª and 9ª Caproni, 31ª and 32ª reconnaissance, 71ª, 72ª, 73ª and 75ª fighters and 1ª Idro. On 12 February it briefly received the 46ª Squadriglia.

On 10 April 1917 the III Gruppo split, forming the new IX Gruppo, remaining with just 1ª Idro, 5ª, 9ª, 31ª, 32ª and 71ª, while *magg.* Ferdinando De Masellis replaced La Polla as commander for just one month, as on 10 May he passed command to *magg.* Adolfo Resio. On that same date the 32ª went to the VII Gruppo, while fighter squadrons were traded with IX Gruppo, getting 72ª and giving 71ª. As the 46ª was permanently attached, the new strength of the III Gruppo was now 1ª Idro, 5ª, 9ª, 31ª, 46ª, 72ª and the new 50ª.

All through the summer, at the command of Resio, now *tenente colonnello*, the III Gruppo of Verona for the 1ª Armata was stable. On 20 August it received the 37ª Squadriglia

Above: Officers of the IV Gruppo with the first American pilots to cross the Italian front. Bonazzi is sitting in the center, behind him, Lt. Le Roy Kiley is identified.

coming from the IX Gruppo, to which it returned in December 1917. The 3ª Idro, to be formed, was assigned to it like the new 61ª replacing the disbanded 50ª. Then started a continuous shuttle of squadrons to and from between III and IX Gruppo.

In October 1917 the III Gruppo had Squadriglie 5ª, 9ª, 31ª, 50ª, 61ª (since 7 Oct.) 120ª (formerly 72ª SAML) 1ª Idro and 3ª Idro but on the 15th its two Caproni units, 5ª e 9ª, deployed at Campoformido, went to the Raggruppamento Squadriglie da Bombardamento, the 31ª and 61ª, went to the IX Gruppo. Finally the 1ª and 3ª Idro were allocated to the Regia Marina and broken down into flights. On 22 October came 75ª and 121ª from the IX Gruppo. The 75ª Squadriglia generated the new 72ª Nieuport also assigned to the III Gruppo, that in November had on strength 37ª, 50ª, 72ª, 75ª, 120ª and 121ª. Then it received one of the three flights of the new 134ª, based at Castenedolo, then on 6 November this squadron moved to Verona passing wholly to the III Gruppo. Five days later the 50ª Farman ceased to exist. *T.col.* Resio went to the *Comando di Aeronautica* of the 1ª Armata and *cap.* Arturo Oddo became commander of the III Gruppo.

On 30 November 1917, the III Gruppo at Verona had the squadrons 75ª, 121ª and 134ª at Verona, 37ª at Castenedolo and 71 ª at Sovizzo. For a few days, until 16 December, the 71ª joined the group, that on 10 December also got from the IX the 31ª at Sovizzo, and five days later the 61ª, that moved from Ghedi to Verona, while giving to IX Gruppo the 37ª.

When the XVI Gruppo was formed, the 121ª moved from Verona to Sovizzo in the new unit together with the 71ª, thus the III Gruppo was left with 61ª, 75ª and 134ª (Verona flight). Until March 1918 the III Gruppo remained with these three squadrons plus a Sezione Farman Difesa of Verona, then on 26 March it received the 135ª Pomilio and the 1ª Sezione SVA, and gave the 61ª to the XVI Gruppo.

On 12 May 1918 the 61ª moved from Verona to Ganfardine and back into the III Gruppo, together with the 1ª Sezione SVA that had gone to the XVI, while the 135ª went the opposite way, going to Castelgomberto. In May therefore the strength of the III Gruppo at Verona consists in 75ª, 61ª, 134ª, 1ª Sezione SVA and Sezione Difesa di Verona, now equipped with SP.

On 22 June *cap.* Arturo Clerici Bagozzi became acting commander. On 6 July the command of the Gruppo moved from Verona to Ganfardine. *Magg.* Gino Matteucci became commander on 15 July. On 4 September the group retained the Squadriglia 134 A and on 28 September it got from the XX Gruppo the 2ª Sezione Idro FBA. On 13 October command was assumed by *magg.* Luigi Biagini.

Above: The crowded airfield at Casoni in December 1917, the base for Pomilios of the II Gruppo and Hanriots of the VI Gruppo.

On Victory Day the III Gruppo was at Ganfardine with 61ª, 75ª, 134ª, 1ª Sezione SVA, 1ª and 2ª Sezione Idro.

On 4 December 1918 the III Gruppo moved back from Ganfardine to Verona with its 75ª. It took 60ª SVA and 3ª Sezione SVA, renamed 1ª Sezione. In December the group was at Gardolo, near Trento, *magg.* Biagini being its commander, with 75ª and 1ª Sezione SVA. In 1919 two new units joined the III Gruppo, 58ª and 59ª SVA. The III Gruppo was disbanded on 1 August 1919.

Gruppo Comando Supremo

The Group of the Supreme Command was formed on 1 November 1915 with Squadriglie da Offesa 1ª, 2ª, 3ª, 4ª at Pordenone. It was the tool to assign the bomber force to strategic objectives, a forward-looking arrangement. Its commander was the chief of the Battaglione Squadriglie Aviatori (the heavier-than-air component of the Corpo Aeronautico Militare) himself, *t.col.*. Barbieri. It then received Squadriglie 6ª and 7ª da Offesa. On 18 February 1916 during the action against Lubiana (Laibach) *t.col.* Barbieri was killed, and *magg.* Gaviglio replaced him. In April 1916, as units were re-organised, the Gruppo Comando became IV Gruppo.

Gruppo Squadriglie di Aviazione per l'Artiglieria

This unit was formed in late June 1915, as Gruppo Squadriglie Aviatori di Artiglieria, (group of squadrons of artillery) with *magg.* Amedeo de Siebert as commander, based at Medeuzza, with the Squadriglie da Artiglieria 1ª Caudron and 2ª Macchi Parasol. On 24 September it received the 3ª Squadriglia Parasol. From Medeuzza it moved to Oleis. In October the new Squadriglie 4ª and 5ª Caudron were added. On 5 November 1915 it took the name of Gruppo di Aviazione per l'Artiglieria.

In March 1916 the Gruppo got 9ª Farman that was re-designated 7ª Artiglieria, deployed at Oleis, while *maggiore* Emilio Gamerra took command of the Gruppo.

The Squadriglie 1ª, 2ª, 3ª, 4ª, 5ª, 6ª, and 7ª Artiglieria on 15 April 1916 became 41ª, 42ª, 43ª, 44ª, 45ª, 46ª and 47ª. On 30 April 1916 the Gruppo became Comando di Aviazione per l'Artiglieria, with commander *t.col.* Amedeo De Siebert, to which reported the new Gruppi V, VI, and VII. This Command was eliminated on 8 April 1917, and its Gruppi, V, VI, and VII lost the designation of units for the artillery.

IV Gruppo

This unit was born out of the *Gruppo Comando Battaglione Aviatori a disposizione del Comando Supremo* according to the new structure of the air force introduced on 15 April 1916. It was based at Pordenone, its commander was *t.col.* Lelio Gaviglio and it included 1ª, 2ª, 3ª, 6ª, 7ª and 8ª Squadriglia Caproni. *T.col.* Alberto Marenco Marenghi then became commander. In June the IV Gruppo received the 10ª Squadriglia, coming from the III Gruppo, in October 1916 the new 13ª Squadriglia, and in January 1917 the new 14ª.

On 13 February 1917 *magg.* Luigi Falchi became commander. On 10 April 1917 the IV Gruppo gave part of its squadrons to form the new XI Gruppo, remaining with 1ª, 8ª, 10ª, 13ª and 14ª. In May the base of the command moved to La Comina airport, and it got the local *Sezione Difesa*, with Farmans. In June also the 5ª Squadriglia was added. In the summer, *magg.* Falchi was removed, being held responsible for the accidental bombing of Italian troops. *Cap.* Cesare Darby was appointed as acting commander. On 1 October 1917 *magg.* Fernando Bonazzi replaced Darby as commander, three days later the Gruppo got 15ª Squadriglia and during the month it lost the 5ª.

The Italian retreat at Caporetto affected the organization of the bomber groups, with confused and hectic transfers of squadrons over a few days, between the IV and XI Gruppo. When in December the situation stabilized, the IV Gruppo, now based at San Pelagio, near Padua, had returned basically to its original structure, with 1ª, 8ª, 13ª, 14ª and 15ª Squadriglia, but the last two left the group on Christmas Eve, destined to the French front, as part of the new XVIII Gruppo.

In February 1918 the 5ª Squadriglia returned. In July the 91ª Squadriglia "Baracca" was attached to the IV Gruppo, and remained until September. In that month another crack unit arrived, the 87ª Squadriglia "Serenissima". In October the IV Gruppo became part of the *Massa da Bombardamento* for the coming final offensive.

At the end of hostilities this Gruppo still had the same squadrons that were its strength since August 1917: 1ª, 5ª, 8ª and 13ª Squadriglia Caproni. By 28 February 1919 it had on charge 1ª, 2ª, 5ª and 8ª, in the spring 1919, when its commander was *magg.* Salvatore Russi, it had 1ª, 5ª and 8ª. In May it included 14ª, re-formed with Caproni Ca. 5, 1ª, 4ª and 8ª. The IV Gruppo was disbanded on 1 August 1919.

V Gruppo

The V Gruppo was born on 30 April 1916 from the Gruppo Aviazione Artiglieria, from which it inherited the Squadriglie 41ª, 42ª, 43ª and 44ª. Its commander was *magg.* Emilio Gamerra, it was based at Chiasottis and assigned to the 3ª Armata, engaged primarily in artillery spotting in the Carso area.

On 10 June 1916 it lost 41ª Squadriglia, that went first to the VII and then to the VI Gruppo, so it retained the 42ª, 43ª, and 44ª for almost one year. In April 1917 *maggiore* Mario Santi became the new commander.

On 5 May 1917 the V Gruppo received the 38ª and 39ª Squadriglia SP from the I Gruppo. On 15 July it got the new 28ª SP, being then formed by 28ª, 38ª, 39ª, 42ª, 43ª and 44ª. In October it received the new 118ª SAML while the Caudron-equipped squadrons, 42ª, 43ª and 44ª were disbanded.

After the Caporetto retreat the V Gruppo included Squadriglie 28ª, 38ª, 39ª and 118ª based on the fields of Marcon and Tessera, near Venice, operating along the Piave front, from Ponte della Priula to the sea. On 7 November it took over at Marcon also the 23ª Squadriglia, coming from I Gruppo. *T.col.* Mario Santi died on 13 November 1917, killed by an enemy bullet as he was flying at very low level against enemy troops over the Piave River.

From 16 December 1917 to 14 January 1918 *cap.* Ottorino Mutti was temporary commander of the group. The V Gruppo was severely taxed, because squadrons were down to two aircraft, having lost all their reserves during the retreat, and the new airplane that was supplied to replace the SP.3, the SIA 7B, was experiencing very serious problems. In order to spot for the artillery of the 3ª Armata even the L.3 flying boats of the Regia Marina from the Stazione Miraglia at Venice had to be used.

In March 1918 the V Gruppo, based at Carpenedo, with *cap.* Ercole Messi as commander, had on strength the 23ª SP.3, 28ª SIA, 118ª SAML, and the 1ª Sezione of the 24ª at Marcon, 38ª SIA, and 39ª SAML at Ca'Tessera. By mid-March it received the 62ª SP.4, while the 118ª went to the XIX Gruppo. The 23ª Squadriglia was assigned to the 5ª Armata and did not report to the group for operations. The crisis was getting worse, because the problems of the SIA 7b could not be solved at squadron level. On 31 May the V Gruppo got the 131ª Pomilio.

During the summer the V Gruppo took over the *Servizio Speciale* for the I Gruppo, that is to say the aircraft used for the "Giovane Italia" intelligence operation that brought or dropped agents behind the enemy lines. On 7 July the 131ª returned to the I Gruppo.

In mid-August Messi left the V Gruppo and took command of the XVI Gruppo. On Armistice Day the V Gruppo was at Carpenedo formed by Squadriglie 28ª, 38ª and 39ª, with *cap.* Pietro Pinna as acting commander. In December it was at Aiello with 38ª and 39ª with *cap.* Vittorio Giovine as commander. It kept this structure until 1 August 1919 when it was disbanded.

VI Gruppo

The VI Gruppo was born as a new artillery group of the 2ª Armata on 30 April 1916 with Squadriglie 45ª and 47ª, based at Oleis, and its commander was *magg.* Mario Santi. On 16 August 1916 it took over also the 41ª coming from the V Gruppo, and kept this strength until 14 November 1916, when the 47ª Squadriglia left, moving to Macedonia. During 1916 its units flew a total of 1,068 war missions and took 600 photographs.

In the spring 1917 *magg.* Gino Zanuso replaced Santi. On 10 April 1917 the VI Gruppo gave its 41ª Squadriglia to the II Gruppo, and in early May it received the 24ª and 29ª that formed a Sottogruppo "Cavazzo" (a town in the Carnia, a region in the north-east of the front) formed on 2 May. It operated side by side with the II Gruppo, also assigned to the 2ª Armata. The VI Gruppo also took on strength the 2ª Sezione of the 113ª and on 9 May also the 41ª. Despite the bad weather, it flew 135 war missions during the offensive of May 1917. In June it consisted of 24ª, 29ª, 41ª, 45ª, 3ª Sezione of the 83ª Nieuport and 2ª Sezione of the 113ª.

The VI Gruppo was much reinforced: on 9 July 1917 it received two fighter units, 76ª and 81ª, and on 25 July the new 27ª SP was ordered to report to it for operational deployment. Therefore in the summer 1917 it had 24ª, 29ª, 41ª, 45ª, 76ª, 81ª, 3ª Sezione 83ª and two Sezioni of 113ª, plus 27ª of the II Gruppo, 24ª, 29ª, 83ª and 113ª forming the Gruppo Cavazzo. The 78ª was briefly assigned to VI Gruppo, from 9 July to 18 August, when it returned to X Gruppo.

On 10 August the VI Gruppo received the new 35ª SP and the new 36ª, while in the autumn 29ª, 41ª, and 45ª were disbanded. Its commander now was *magg.* Eginardo Bongiovanni.

On 2 November 1917, after the retreat, the VI Gruppo was disbanded with the formation of a *sottogruppo* (subgroup) with just the artillery squadrons, 24ª, 27ª, 35ª, 36ª and 39ª and a sottogruppo with the fighter squadrons 76ª, 78ª and 81ª that reported to *t.col.* Piccio who took the command of all the fighters of the different Aeronautiche d'Armata (air corps of the Armies).

The VI Gruppo was reformed on 10 November 1917 with the 76ª, 78ª and 81ª Squadriglia da caccia; its commander was *magg.* Salvatore Calori and it was assigned to the new *Comando d'Aeronautica della 4ª Armata*, at Istrana. The Gruppo was the protagonist of the so-called Battle of Istrana, on 26 December 1917, and for this feat it was awarded the *Medaglia di Bronzo al Valor Militare*, the only aviation unit decorated in World War 1. From its rebirth to the end of 1917 the group flew 273 war flights.

On 23 January 1918 the 78ª Squadriglia went to the new XV Gruppo. On 2 February 1918 the VI Gruppo moved from Istrana to Isola di Carturo, as the former airfield was wrecked by continuous night bombings. It received the 3ª Sezione SVA and in March it moved to Casoni, with the command office at Rossano. On 30 March *magg.* Vincenzo Lombard became commander. This structure didn't change, except for the periods when the 76ª Squadriglia was assigned to the *Massa da Caccia* for the battle of the Piave.

The scoreboard of the VI Gruppo included 626 air combats with 179 victories and 14 aces. Its pilots were awarded two *Medaglie d'Oro*, 36 Silvers, 18 Bronzes and 40 War Crosses. The VI Gruppo was disbanded on 1 August 1919.

VII Gruppo

The VII Gruppo was born as an artillery unit on 30 April 1916, formed by the 46ª Squadriglia Farman at Verona and its detachment at Asiago, assigned to the 1ª Armata on the north-western front. Its commander was *magg.* Giulio Costanzi. In mid-June it acquired the 43ª and then the 42ª, deployed from the Carso for the battles on the plateau of Asiago, based at Trissino since 19 June, and kept them until July, when the Austrian offensive was pushed back. The VII Gruppo moved on 19 June from Verona to Trissino. In July it got the new 49ª and on 29 July it moved from Trissino to Nove.

On 1 January 1917 it was at Nove di Bassano with its 46ª at Castenedolo and Trissino and the 49ª at Nove. On 10 April 1917 it received the 48ª Squadriglia and one Sezione of the 113ª SAML at Feltre. On 10 May 1917 the VII Gruppo was assigned to the new *Comando d'Aeronautica della 6ª Armata*, having command over the 32ª, received from the IX Gruppo, 49ª, 79ª, formerly of the X Gruppo and 26ª in formation, and giving the 48ª to the new XII Gruppo and the 46ª to the III Gruppo.

In the summer 1917 the VII Gruppo had Squadriglie 26ª, 32ª, 49ª, 79ª and 50ª. On 14 September 1917 it received the 33ª from the II Gruppo, and six days later also the new 62ª SP.4 and the 115ª. The *Comando d'Aeronautica della 6ª Armata* was disbanded and in autumn the VII Gruppo, still at Nove but now with *magg.* Amedeo De Sanctis as commander, reported to the *Comando Truppe Altipiani* (Command of the troops of the plateaus) with its Squadriglie 26ª, 32ª, 33ª, 62ª, 79ª and 115ª. Its next commander was *t.col.* Glauco Visconti.

In November the group had on strength 26ª, 32ª, 33ª, 79ª, 115ª Squadriglia and 2ª Sezione of the 139ª. On 10 December 1917 it gave away 79ª, 115ª and 139ª that formed the new XV Gruppo and it remained with 26ª SP.3, 32ª Farman and 33ª SP.3, returning under the command of *cap.* Amedeo De Sanctis. This structure remained in early 1918, then the VII Gruppo received the 3ª Sezione of the 24ª, that joined the other units at Nove di Bassano.

On 4 March 1918 the VII Gruppo returned under the re-formed *Comando di Aeronautica della 6ª Armata*. On 10 March it received a Sezione (flight) of the 83ª Caccia. On 13 March it moved to San Pietro in Gu, in the buildings vacated by the XV Gruppo, with its three Squadriglie, leaving Nove free for the French squadrons. In May it got the 2ª Sezione SVA and there were no further changes to its structure until 4 October, when it gave its 83ª to the new XXIV Gruppo, then on 25 October it received the 136ª Pomilio. During the summer it operated as a flying school, training pilots on the Pomilio PE at San Pietro in Gu, with *ten.* Santerni as chief instructor. On Armistice Day therefore the VII Gruppo was at San Pietro in Gu with 26ª, 32ª, 33ª and 136ª.

In December it had 26ª, 32ª, 33ª and 83ª, while in 1919, after the first de-mobilisation of units, it had 32ª, 83ª and 2ª Sezione SVA. In April 1919, De Sanctis still being the commander, it had 32ª and 118ª both equipped with SVA 10 at Fossalunga. On 1 August 1919 it turned into *Comando di Gruppo da Ricognizione per Artiglieria* (command of

Above: Pilots of the 70ª and 82ª Squadriglia, X Gruppo, at Gazzo in September 1918. All the pilots are identified.

reconnaissance group for artillery) at Aiello with Squadriglie 31ª SVA 10 at Bolzano, 32ª SVA 10 at Campoformido, 38ª PE at Zaule, 39ª PE and R2 at Aiello.

VIII Gruppo

The VIII Gruppo was formed on 9 December 1916 in Albania, based at Krionero (4 kms from Valona), with Squadriglie 11ª Caproni and a Sezione Nieuport, at Tahyraqua, 34ª Farman at Krionero. Its commander was *magg.* Egidio Carta. In August 1917 *cap.* Leopoldo De Rada became commander. On 23 September 1917 the Sezione Nieuport became 85ª Squadriglia Nieuport. The 34ª Farman was disbanded in November 1917, replaced by the 116ª SAML.

This group, serving the Italian XVI Corpo d'Armata of Albania, had no changes until 16 January 1918, when *cap.* Ernesto Pellegrino from the 85ª Squadriglia replaced De Rada.

Formed by the 11ª at Tahyraqua, 85ª at Piskupi and 116ª at Valona, the VIII Gruppo took part in the allied offensive of May 1918. On 16 June 1918 *cap.* Alberto Livi became its temporary commander but, piloting a British Sopwith Camel serialed C45, he crashed in the sea and died on 3 August 1918. He was temporarily replaced by *cap.* Mario Alberto Ponis, commander of the 85ª Squadriglia. The group was actively engaged in the summer offensive that brought the conquest of Berat and Fjeri. On 27 August 1918 *magg.* Aurelio Liotta became commander. The group took part in the September offensive in which Durazzo was conquered, and it immediately deployed there a detached flight formed by fighters and two Pomilio PE.

The operations of the VIII Gruppo in Albania during the war included over 800 war flights, 100 tons of bombs dropped, 80 air combats, one *Medaglia d'Oro*, twelve *Medaglie d'Argento* and 19 Bronzes.

In the spring 1919 the VIII Gruppo kept its 11ª Squadriglia at Tahiraqa, 85ª at Piskupi, 116ª SAML at Valona, one SVA flight at Valona and one PE and HD flight at Durazzo. *Magg.* Liotta was commander of *Aeronautica dell'Albania*. The aircraft of the group were used to set up a mail service from Valona to the other towns of Albania, a country that hardly had any roads. On 10 October 1919 the 85ª Squadriglia Caccia and the 116ª Ricognizione were disbanded: only one flight from each remained, which were merged with the Sezione SVA of Valona to form a Squadriglia Mista. Having the 11ª Caproni, the Squadriglia Mista and the Sezione di Durazzo, the VIII Gruppo, with commanders *cap. oss.* Pasquale Mento, then *cap. oss.* Claudio Coraglia, was engaged in June and July 1920 in combats against Albanian insurgents. Then it was disbanded, as Italian occupation of Albania came to an end.

IX Gruppo

The IX Gruppo was formed on 10 April 1917 splitting the

III Gruppo, with Squadriglie 37ª, 72ª, 73ª, 74ª, and 75ª formerly of the III, and then with the new 112ª. It was based at Villaverla, assigned to the 1ª Armata, northern sector of the front, with *magg.* Giorgio Chiaperotti as commander. On 10 May it gave to the III Gruppo its 72ª receiving in return the 71ª. It also deployedan unattached Sezione Spad at Verona. In July the 112ª left the group, moving to Chiasiellis and going to the I Gruppo. On 28 July *magg.* Carlo Melegari became commander. On 20 August the 37ª Squadriglia went back to the III Gruppo.

From September 1917 the structure of the IX Gruppo nominally changed, as the 73ª and 74ª got a new designation, becoming 121ª and 122ª thus the Group now had 71ª, 75ª, 121ª and 122ª. On 22 October 1917 it lost the 122ª that was assigned to the *Comando Difesa Aerea*. On 24 October the 75ª and 121ª were returned to the III Gruppo, the latter passing to the IX Gruppo 31ª and 61ª SP.4, and also two flights of the new 134ª Pomilio went to the group at Villaverla.

In the period of reorganization after the retreat, there were countless exchanges of units between III and IX Gruppo. Finally, by 20 November 1917, the IX, still at Villaverla with *magg.* Melegari as commander, consisted of 61ª, 71ª and 1ª Sezione 134ª at Castenedolo and 31ª at Castelgomberto. On 24 November *magg.* Chiaperotti returned as commander of the IX Gruppo. On 10 December the IX got from the III the 37ª and 120ª at Castenedolo and gave it 31ª, 61ª and 71ª, now thus having 37ª, 72ª, 120ª and 135ª (a new unit, developed out of the Sezione of the 134ª already assigned to this group) all at Castenedolo.

The group kept this structure until February 1918, when the 135ª moved to Sovizzo going to the XVI Gruppo, while the IX received the 62ª SP.4 and, at the end of the month, also the 113ª that was sent back away from the frontline for training purposes. Since March 1918 thus the IX Gruppo had 37ª, 62ª, 72ª, 120ª and one Sezione Farman di Difesa at Castenedolo, plus the 113ª at Medole. In mid-March it gave its 62ª to the V Gruppo, and briefly controlled the 40ª Squadriglia. When the new XX Gruppo was formed, on 26 March 1918, the IX remained with 120ª, 112ª (arrived on 24 March from the I Gruppo) 72ª, II/83ª and 2ª Sezione SVA. In the spring two new units arrived, the new 136ª Pomilio and the 74ª Caccia.

In May 1918 the IX Gruppo, still at Castenedolo at the command of *magg.* Giorgio Chiaperotti, and assigned to the 1ª Armata, had Squadriglie 72ª, 74ª, 112ª, 113ª, 120ª and 136ª. In late May it received the 6ª Sezione SVA, and gave the 113ª to the XX Gruppo.

In July 1918 the IX Gruppo was assigned to the new *Comando d'Aeronautica* of the 7ª Armata with its 72ª, 74ª, 112ª, 120ª, 136ª and 6ª Sezione SVA with *magg.* Luigi Tagliasacchi as commander. With the reorganization of the units of the 7ª Armata, the IX Gruppo became a unit of *Ricognizione d'armata* (army, that is, strategic, reconnaissance) with 112ª, 120ª and 6ª Sezione SVA, giving away on 18 July 72ª and 74ª Caccia that formed the new XXIV Gruppo, and the 136ª that went to the XX Gruppo, to be joined by the 120ª on 25 August.

Variations of structure were not finished yet: the IX Gruppo on 25 September received the 74ª Caccia from the disbanded XXIV, then on 4 October gave it to the XX. On Armistice Day, 4 November 1918 the IX Gruppo was still at Castenedolo with 112ª and 6ª Sezione SVA. In December it was at Campoformido, with *cap.* Giulio Cesare Paldaoff commander, formed by 74ª and 9ª Sezione SVA. It was disbanded on 6 February 1919.

X Gruppo

The X Gruppo was formed on 10 April 1917 with Squadriglie 33ª, 70ª, 78ª, 79ª and 82ª at Pasian del Prato, close to Campoformido airfield, at the command of *magg.* Pier Ruggero Piccio assigned to the Comando Supremo. Soon after, in early May, it gave the 79ª Squadriglia to the VII Gruppo, and got the new 91ª. In May the 33ª Squadriglia was disbanded, replaced by the new 111ª SAML. In July it got the new 114ª SAML but lost the 111ª that went to Macedonia, and temporarily lost also the 78ª, that returned in August. In mid-September the 114ª went to the II Gruppo, while the 132ª arrived.

In autumn 1917, Squadriglie 70ª, 78ª, 82ª, 91ª and 132ª formed the X Gruppo, still at the command of Piccio. On 25 October the 132ª returned to the II Gruppo, and the X Gruppo was now a unit formed only by fighter squadrons. It abandoned Udine in the course of the retreat, while *magg.* Gino Matteucci became commander. In early 1918 the group was at San Pietro in Gu, with its Squadriglie at different airfields. It received also the 87ª Squadriglia SVA from Ghedi. In late March the command was at Istrana together with 70ª and 82ª, while 87ª was at Ghedi and 91ª at Quinto di Treviso. Also, the 5ª Sezione Difesa di Padova relied on the X Gruppo for administrative chores. At the command of Matteucci, the X Gruppo in the spring moved to Gazzo, where 70ª and 82ª also deployed, and kept this location until the end of the war. In September the 91ª Squadriglia "Baracca" went to the new XVII Gruppo. In December *magg.* Ercole Ercole became its commander, and then the X Gruppo was disbanded, on 10 December 1918.

XI Gruppo

The XI Gruppo was formed at Aviano splitting up the IV Gruppo on 10 April 1917 with 2ª, 3ª, 4ª, 6ª and 15ª Squadriglia Caproni, at the command of *magg.* Piero Opizzi. On 14 July 1917 Opizzi left command of the XI Gruppo to *magg.* Domenico Mondelli, who kept it only eleven days, replaced on 25 July by *magg.* Armando Armani, having been held responsible, like Falchi, for an accidental bombing of Italian troops by a single Caproni.

In the summer 1917 the XI Gruppo distinguished itself in the first bombing of Pola, on 2 August. From its ranks 14 Capronis were drawn to form the Distaccamento A.R. for

Above: The officers of the command of the XIX Gruppo. In the center is *cap*. Napoleone Del Duca.

the bombing of Cattaro. On 4 October the XI Gruppo gave its 15ª Squadriglia to the IV Gruppo receiving from it the 5ª on 27 October, when the retreat had begun. After Caporetto, the XI Gruppo moved to Padua and sent some of its units to Ghedi for reorganization. In November the group gave half of its squadrons to form the XIV Gruppo and was left with the 4ª, 5ª and 6ª.

On 6 January 1918 *magg*. Armando Armani became chief of the *Raggruppamento Squadriglie da Bombardamento* (bomber squadrons wing). On 3 February the 5ª Caproni returned to the IV Gruppo while on the 21st a new unit was attached, the "Serenissima" 87ª, as *Squadriglia Autonoma.* In the spring 1918 commander of the group was *magg.* Leonardo Nardi. In March the XI Gruppo was still divided between the airfields of Verona and Ghedi and it got the 181ª Squadriglia Triplani of *magg.* Carnevale.

On 19 May the XI Gruppo moved to Ca degli Oppi, some 20 km from Verona. The next month *magg.* Luigi Reggio became its commander and the group moved back to Verona, closer to the front, to combat the Austrian offensive. In July the 181ª went to the XXII Gruppo and 4ª, 6ª and 87ª then formed the XI Gruppo. In September also the 87ª left the group and joined the IV.

On 7 October *magg.* Reggio died, victim of the "Spanish influenza" that was to kill several Italian airmen. Command of the XI Gruppo was temporarily entrusted to *magg.* Enrico Cellerino. In October the group became part of the *Massa da Bombardamento* for the coming final offensive. On 20 October the XI Gruppo moved to Padua leaving one airplane at Verona for the Ufficio ITO (intelligence service) of the 3ª Armata.

In December *magg.* Costantino Quaglia was commander of the XI Gruppo that at the end of the year moved to Verona with its Squadriglie. On 28 February 1919 these were 4ª, 6ª and 161ª SIA 9b.

XII Gruppo

The XII Gruppo was formed, like so many other units, on 10 May 1917, assigned to the new *Comando di Aeronautica* of the 4ª Armata. It included the 48ª Squadriglia Caudron, 2ª Sezione Nieuport of the 83ª Squadriglia, 35ª Squadriglia SP that was about to be formed (that however went to the VI Gruppo) and the 3ª Sezione of 113ª. It was based at Belluno, with commander *cap.* Felice Porro, who in June left the command to *magg.* Novellis. In the summer Porro was commander again. In August the group received the 115ª SAML that incorporated the flight of the 113ª at Villaverla.

The *Comando di Aeronautica* of the 4ª Armata was disbanded but the XII Gruppo remained available to that

Army, and got back the 35ª SP.3. On 3 October it received the new 117ª SAML, then on the 15th the 35ª SP went back again to the VI Gruppo. On 3 November the 27ª and 22ª of the II Gruppo, retreating from the advancing enemy, were attached to the XII Gruppo.

On 10 November 1917 the *Comando d'Aeronautica* of the 4ª Armata was reactivated, obviously getting the XII Gruppo of *cap.* Porro. The group received the 36ª of Istrana, transferred its 117ª to the I Gruppo and so was formed by the Squadriglie 22ª and 36ª at Istrana, 27ª, 35ª, and 48ª at Castello di Godego.

On 15 January 1918 *cap.* Mario Martucci became commander. In the spring the XII Gruppo was based at Cittadella with its 22ª SIA, 36ª SIA and the 2ª Sezione of the 24ª at Casoni, 27ª and 35ª SIA and 48ª Caudron G.4 at Castello di Godego, and this structure remained until July 1918.

When many SIA Squadriglie were disbanded, the XII Gruppo was left with the 22a and 36a.The 24a and 35a were disbanded while the 27a and 48a were used to reform the II Gruppo at Castello di Godego. This Gruppo had been left without any units due to the disbandment of the 132a and 133a. Commander Martucci went to the XX Gruppo, and *cap.* Ettore Lodi progressed from the 22ª to the command of the Gruppo on 16 July 1918. Then this unit, based at Casoni, had no further change until the end of the war.

In December, still with commander Lodi and consisting of 22ª and 26ª, it was at La Comina. On 28 February 1919 it had on charge 21ª, a new 24ª was in the process of formation, and 4ª Sezione SVA. In April this group was at Casoni, with 21ª and 36ª SAML. Presumably it was disbanded at the beginning of summer 1919.

XIII Gruppo

The XIII Gruppo was formed on 10 November 1917, uniting fighter Squadriglie 77ª, 80ª and 83ª, at the service of the 3ª Armata, with *cap.* Mario Gordesco commander. In November it had 77ª, 80ª, 2ª Sezione of the 83ª and 84ª at Marcon. When the 84ª was disbanded, the group included 77ª Spad, 80ª Hanriot and 83ª Nieuport.

On 4 March 1918 it lost the 83ª Squadriglia. On 8 June 1918 command passed to *cap.* Umberto Venanzi. This unpopular commander was replaced on 17 September by *cap.* Alberto De Bernardi. On the eve of the final battle the 77ª Squadriglia joined the *Massa da Caccia*, and the XIII Gruppo remained with only the 80ª. In November it was composed of only the 77ª Squadriglia and a seaplane squadron, in December it was disbanded, then at the end of the month the 77ª returned, the group was briefly reactivated and then it was definitely abolished.

XIV Gruppo

This unit was originally formed as Gruppo Squadriglie di Ghedi, with the Caproni units retreating in November 1917 and it got the designation XIV Gruppo on 14 November 1917, as part of the *Raggruppamento da Bombardamento per il Comando Supremo. Magg.* Vittorio De Muro was its commander and it had the 2ª, 3ª, 7ª, 9ª and 10ª Squadriglia. In early December this group moved its base to Padua.

On Christmas Eve 1917 the 3ª Squadriglia was assigned to the XVIII Gruppo for service on the French front. On that same day the Gruppo got a new commander, *magg.* Oreste Salomone, the hero of the Lubiana raid of 18 Feb. 1916. On 2 February 1918 Salomone was killed in a crash landing a Caproni of the 10ª Squadriglia, after the first moonless night bombing. The group's command was temporarily passed to *cap.* Leone Cola De Renzis, commander of the 2ª Squadriglia, until the arrival of *cap.* Diego Sabbatini. Then on 10 April *magg.* Francesco Vece became commander.

On 14 April 1918 the XIV Gruppo took over the 16ª Squadriglia, but his unit was disbanded on 20 June. On 23 July 1918 the group moved from Padua to Arquà.

On 15 September the 9ª Squadriglia left the XIV and was assigned to the IV Gruppo. On 23 September *magg.* Salvatore Russi replaced Vece as commander. In October the Gruppo became part of the *Massa da Bombardamento* with its three units, 2ª, 7ª and 10ª Squadriglia. The Gruppo was disbanded on 28 February 1919.

XV Gruppo

The XV Gruppo was formed on 15 December 1917 with 115ª SAML, 139ª Pomilio and 79ª Nieuport, three squadrons all at San Pietro in Gu, with *t.col.* Glauco Visconti as commander. Right after, the group received the 2ª Sezione SVA of the 86ª Squadriglia. On 24 January 1918 it also received the 78ª Hanriot formerly of the VI Gruppo. In March 1918 it was at San Pietro in Gu with 78ª and 79ª Caccia and 2ª Sezione SVA at Nove, 115ª SAML and 139ª Pomilio at San Pietro in Gu. On 18 March the command of the XV Gruppo moved to Paese and was assigned to the newly reformed *Comando di Aeronautica* of the 2ª Armata.

In mid-March 1918 the Squadriglie moved to San Luca, near Istrana, leaving San Pietro in Gu to the units of the VII Gruppo that left Nove to the French squadrons, while the 2ª Sezione SVA was assigned to the IX Gruppo, being replaced in April by the new 4ª Sezione SVA, attached first to the 78ª and then to the 115ª Squadriglia.

In the summer the XV Gruppo was ordered to report to the new 8ª Armata. In July it transferred its fighter units, 78ª and 79ª, to the new XXIII Gruppo, as the *Comando Superiore* now favored single-role groups, it lost also the 139ª that was disbanded and it remained with just 115ª SAML of *magg.* Pascale and 4ª Sezione SVA of *ten.* Persegani. These two units were tasked with infantry service for the 8ª Armata (Army Corps VIII, XXII and XXVII). Commander since 18 July was *magg.* Enzo Baroni. Then the XV Gruppo got on 12 October the new 50ª Squadriglia SVA that on 24 October merged with the 4ª Sezione SVA becoming 56ª Squadriglia.

In the post-war period the XV Gruppo initially had Squadriglie SVA 56ª, 58ª, 59ª and 60ª. In December 1918

Above: Another souvenir photo of Italian airmen of the XIV Gruppo with five American pilots. Sergeants mixing informally with officers, including a Major, would have raised eyebrows in Britain. The first American pilot on the right is wearing both the US wings on his chest and the Italian eagle on his sleeve.

it had 56ª, 57ª, 58ª and 60ª at Ganfardine, 59ª and 62ª at Montichiari, with *magg.* Enzo Baroni commander. The plan was to organize it with four Squadriglie, two of SVA single-seaters and two of SVA two-seaters, that is 56ª, 58ª, 59ª and 60ª, but on 6 February 1919 58ª and 59ª went to the III Gruppo and the XV with 58ª and 59ª moved to Busiago. The XV became *Gruppo da Bombardamento Leggero* (Light Bomber Group) and on 28 February it had on strength 56ª, 60ª and 87ª. It then moved to Verona, at the command of *cap.* Luigi Sella. The XV Gruppo was disbanded on 1 August 1919.

XVI Gruppo

The XVI Gruppo was formed at Sovizzo, on the front north of Vicenza, on 16 December 1917 with Squadriglie 31ª, 71ª and 121ª all at Sovizzo for the *Comando di Aeronautica* of the 1ª Armata; its temporary commander was *cap. pil.* Amerigo Notari, commander of the 71ª. In late December it received also the 1ª Sezione SVA. On 4 January 1918 *cap.* Giulio Palma di Cesnola was appointed as temporary commander.

On 13 February the XVI Gruppo received the 135ª Pomilio coming from the IX Gruppo. From 26 March to May 1918 it also had in charge the 61ª SP.4 and SIA coming from the III Gruppo. In early April *cap.* Franco Scarioni, leader of the 31ª Squadriglia, became temporary commander of the group. On 12 May the 1ª Sezione SVA went to the III Gruppo and the XVI Gruppo moved from Sovizzo to Castelgomberto together with its Squadriglie, 31ª, 71ª, 121ª and 135ª.

On 29 May, Palma di Cesnola returned as actual commander of the XVI Gruppo. On 12 July the 135ª Squadriglia was disbanded. On 14 August *cap.* Ercole Messi, coming from the V Gruppo, was appointed commander of the XVI, which on 4 September received the Squadriglia 134 B. On 23 September the 71ª Spad was assigned to the new fighter XVII Gruppo.

In December the XVI Gruppo was at recently conquered Bolzano with 31ª, 121ª and 134ª, still with *cap.* Messi. By 28 February it had 31ª and 121ª, and it was disbanded on 1 August 1919.

XVII Gruppo

The XVII Gruppo was formed on 23 September 1918 with the 71ª Squadriglia Spad coming from the XVI and the 91ª "Baracca", coming from the IV, at Quinto di Treviso, assigned to the Comando Supremo. On 16 October it received also the 72ª Squadriglia Hanriot. Its temporary commander was *cap.* Fulco Ruffo di Calabria.

After the war it lost the 72ª and remained at Quinto with 71ª and 91ª at least until February 1919, soon after it was disbanded.

XVIII Gruppo

In late 1917 France announced its wish to have a force of Caproni bombers on the Western front. The Italian Comando Supremo ordered on 24 December the formation at Taliedo (Milan, where the Caproni factory was located) of the XVIII Gruppo, calling from the front the 3ª Squadriglia, of the XIV Gruppo, and 14ª and 15ª of the IV. The new unit, including 27 pilots, 12 observers and 17 gunners, had as commander *capitano* Renato De Riso.

The XVIII Gruppo left Italy on 9 January 1918 reaching by train Longvic airfield, near Dijon, where its airplanes, part flown and part ferried by rail from Italy, were fitted. These originally were Capronis 2363, 2366, 2378, 4053, 4086, 4219, 4225, 4229, 4232, 4235, 4336.

On 19 February the Gruppo was ready and it went to the front, flying to Ochey airfield, while Longvic was used as a supply and ferry base for the bombers coming from Italy. The Italian group was attached to the *Escadre 11* together with the French *Groupes de Bombardement* 1, 2 and 7, commander Chabert, assigned to the Aeronautics Command of Armies Group East.

Despite the negative weather, the XVIII Gruppo was very active attacking the enemy rear. From February to 3 April its aircraft dropped 55.000 kgs of bombs over enemy installations at Metz, Sablon, Thionville, Couflens, Saarebruch, Laon, Briery and S. Quentin. In the night of 22-23 February, the Italians dropped 5.500 kgs on factories at Ludwigshafen flying in the fog and with a wind of 7 m/s. On 2 April the Gruppo moved to Villeneuve airfield.

On 10 June 1918 the 14ª Squadriglia was ordered to return to Italy to convert on the Caproni Ca. 5 leaving its equipment to reinforce the other two units. On 7 August the Gruppo moved to Chernizey airfield. On 14 October the 15ª Squadriglia returned to Italy also to re-equip on the Ca.5, but six days later the Gruppo got back the 14ª, that had completed its conversion to the new bomber, and that had eight of them on strength.

The end of the war arrived when the XVIII was at Chernizey with its 3ª and 14ª Squadriglia, and it was disbanded on 4 February 1919. The Capronis flew 110 bombing missions on the Western front, its airmen were awarded four *Legion d'Honneur*, two *Medaille Militaire* and ten *Citations a l'ordre de l'Armée* (French awards) plus forty Italian Silver and Bronze medals.

XIX Gruppo

The XIX Gruppo was formed on 17 March 1918 with Squadriglie 114ª and 118ª and the 3ª Sezione of 24ª SIA, these units moving to Istrana, while the Gruppo established its command at Falzè. It reported to the *Comando d'Aeronautica* of the 2ª Armata and its temporary commander was *cap.* Ottorino Mutti. The Gruppo on 25 May received the 23ª Squadriglia that moved from Medole to Istrana.

Cdr. Mutti died in action on 23 June 1918, and command of the Gruppo was given to *cap.* Napoleone Del Duca. Meanwhile the XIX Gruppo had passed from reporting to the abolished 2ª Armata to reporting to the new 8ª Armata, holding the crucial Montello sector. In the first week of July the III/24ª was disbanded, and the XIX became *Gruppo per il servizio di Artiglieria* (group for artillery service) of the 8ª Armata with its three Pomilio Squadriglie: 23ª of *cap.* Simoncelli for the XXII C.d'A., 114ª of *cap.* Fossombrone for the XXVII C.d'A. and 118ª of *cap.* Bardella for the VIII C.d'A.

The group maintained this structure until victory, and in 1919 it moved its command to Fossalunga. On 28 February 1919, when it ceased existing, it was formed by 115ª, that had joined it on 30 November, and 118ª, the 114ª having been disbanded earlier in the month.

XX Gruppo

The XX Gruppo was formed on 26 March 1918 at Castenedolo, under the command of *magg.* Giorgio Chiaperotti, but initially under the temporary command of *cap. oss.* Mario Lanciani, with Squadriglie 37ª SP.3 and 40ª SIA, and to be assigned a third SIA unit in formation, for the new *Comando di Aeronautica* of the 7ª Armata, on the northern front. The situation didn't change until June, when the group instead received the 113ª SAML, formerly of the IX Gruppo, based at Medole and then moving to Cividate Camuno.

In early July the 37ª and 40ª Squadriglia went back to the Centri Formazioni Squadriglie to be disbanded, and the XX Gruppo was reformed on 10 July as a *Corpo d'Armata* unit for the 7ª Armata with the 113ª SAML and 136ª Pomilio, at Brescia, (Castenedolo) with commander *cap.* Mario Martucci, then *magg.* Luigi Tagliasacchi. In August 1918 the group moved to Cividate Camuno for the offensive on the Ortles Mountain. It received on 1 August the 2ª Sezione FBA of the Lake of Iseo and on 25 August the 120ª Pomilio that moved from Medole to San Marco. Then the group also received a flight of the 74ª Caccia that moved to the field of Cividate Camuno. The XX Gruppo was active during the attack on the Tonale Mountain, 12 to 14 August, with reconnaissance and ground attack missions.

On 28 September the 2ª Sezione FBA went to the III Gruppo, while the XX Gruppo on 4 October got the 72ª and 74ª Caccia from the IX Gruppo. On Armistice Day the group was at Brescia with 74ª, 113ª, 120ª and 136ª, but post-war it had 112ª, 113ª and 136ª, as the 120ª was disbanded

Above: A leaflet dropped on the Austrian lines, showing the good treatment of prisoners. (Archive Longhi)

on 1 December. During that month the XX Gruppo took its three Squadriglie to Campoformido, with *magg.* Tagliasacchi commander. On 18 January 1919 it was the turn of the 112ª to be disbanded. On 6 February 1919 the XX Gruppo received the 9ª Sezione SVA. In March it reformed for the 8ª Armata with 113ª, 115ª and 8ª Sezione SVA.

In April its commander was *magg.* Juretigh. On 1 August 1919 it became *Comando di Gruppo da Ricognizione Tattica e Strategica* at Campoformido, with Squadriglie 113ª and 115ª SAML at Campoformido, 121ª SAML at Bolzano, 87ª SVA 5 at Aiello and 58ª SVA 10 at Verona.

XXI Gruppo

The XXI Gruppo was formed at Thessaloniki, Greece, on 25 May 1918 with the Squadriglie operating on the front of Macedonia: 111ª SAML and 73ª Nieuport, with *cap.* Mario Ajmone Cat as temporary commander, to whom reported also one squadron and one flight of the French air force. The XXI Gruppo performed 600 war flights, including 145 bombings, and took 3,000 photographs. The group returned to Italy and was disbanded on 15 August 1919.

XXII Gruppo

The XXII Gruppo was formed on 5 July 1918 with Squadriglie 181ª and 182ª Caproni 4 triplanes at Ghedi, with commander *magg.* Costantino Quaglia. Initially this group was assigned to a *Raggruppamento Squadriglie da Bombardamento*, (bomber squadrons wing) that was disbanded on 31 July so the XXII Gruppo reported to the *Comando d'Aeronautica a Disposizione* (the strategic force) of the Comando Supremo.

On 9 September Squadriglie 181ª and 182ª were assigned to the Regia Marina, and they moved to Poggio Renatico, so the command of the XXII Gruppo was abolished on 20 September.

Another XXII Gruppo was formed on 20 October as a SVA group at the command of *cap.* Francesco Forquet with 89ª and 90ª SVA, at Busiago, north of Padua, assigned to the *Massa da Bombardamento*. This unit took active part in the Battle of Vittorio Veneto attacking bases, railroads and troops. On 27 October alone the group flew 30 missions against the enemy.

After the victory the XXII Gruppo was destined to Libya. In January 1919 it became *Gruppo da Bombardamento Leggero*, (light bomber group) but presumably it was abolished in early 1919, when the 89ª and 90ª moved to the North African colony.

XXIII Gruppo

The XXIII Gruppo was formed on 8 July 1918 with Squadriglie 78ª Hanriot of *cap.* Riva and 79ª Nieuport 27 of *cap.* Mazzini, its commander was *cap.* Antonio Bosio. Assigned to the 8ª Armata, it was based at San Luca and this situation didn't change until the end of the war. The group flew 881 war missions, had 39 air combats, claimed ten victories, suffered three pilots killed and earned 38 decorations.

In December 1918, still with Bosio as commander, the XXIII Gruppo was at Fossalunga. It was disbanded on 1 August 1919.

XXIV Gruppo

The XXIV Gruppo was formed on 18 July 1918, at the command of *magg.* Giorgio Chiaperotti, at Castenedolo, composed by fighter Squadriglie 72ª and 74ª, assigned to the Comando di Aeronautica of the 7ª and 9ª Armata. On 10 August *cap.* Giulio Palma di Cesnola became commander. The XXIV Gruppo was disbanded on 25 September, passing its 72ª to the XX Gruppo and its 74ª to the IX Gruppo.

Another XXIV Gruppo was formed on 4 October 1918 for the 6ª Armata with the 83ª Squadriglia Hanriot and the 2ª Sezione SVA, its command was given to *cap.* Giulio Cesare Paldaoff. This group was based at Poianella and it was disbanded after the victory.

Raggruppamento Squadriglie da Bombardamento

This Command, directly reporting to the Supreme Command, had control of the bomber units of the IV and XI Gruppo and the other units with a strategic mission, such as the 87ª Squadriglia. Its commander was *t.col.* Armando Armani.

Gruppo Speciale Aviazione I

The Gruppo Speciale Informazioni (intelligence) was formed on 5 September 1918 with the 9ª Squadriglia Caproni, with four Ca 450 hp, the Sezione Voisin and the 1ª Sezione SVA, plus one special SP.4, at the command of *magg.* Luigi Carnevale, reporting to the *Comando d'Aeronautica* of the 3ª Armata. Its base was Ca'Tessera and its task was to drop or land agents behind the enemy lines, to relay their messages and to supply them with air drops in pre-set localities.
By Armistice Day it included the 9ª Caproni and one Squadriglia Mista, formed on 1 October. It was disbanded on 21 November 1918.

Gruppi Difesa Aerea

On 14 September 1916 territorial air defence groups were formed. They were *Gruppo di Difesa antiaerea* 1° *Settentrionale*, (northern) with *magg.* Resio commander, with the squadrons and flights located in northern Italy outside of the Zona di Guerra (war zone), and *2° Meridionale*, (central and southern Italy) of *magg.* De Masellis. In October 1917 the air defense of the national territory, reporting to the Supreme Command, was turned over to the *Ufficio Difesa Aerea* of the *Direzione Generale di Aeronautica* of the Ministry of Weapons and Munitions, since these were not units in the "*zona di guerra*", the war zone, where the Supreme command ruled, but they operated in "*paese*", the rest of the country, where the Government ruled. Consequently, the Squadriglie and Sezioni di Difesa Aerea were assigned to report to the Gruppi di Difesa Aerea. There were three Gruppi Aviazione D.A., Northern, Central and Southern, based respectively at Milan, Bologna and Rome. The central and Southern Gruppi were disbanded on 1 March 1918, with the goal of reducing personnel, their units since then reporting directly to the *Ufficio Difesa Aerea*.

Gruppo Aviazione da Difesa Aerea Settentrionale

The Gruppo Difesa Aerea per l'Italia Settentrionale (northern) was formed in September 1916 with commander *magg.* Resio, based at Milan. By 1 January 1918 at the command of *magg.* Ercole, it had as dependent units the 122ª Squadriglia and the Sezioni Difesa of Piacenza, Ponte San Pietro, Taliedo, Cameri, Malpensa and Mirafiori, all in Northern Italy. In January 1918 it was replaced by the Gruppo Difesa Aerea Lombardia, commanded by *t.col.* Zanuso, with Squadriglie 108ª and 109ª at Trenno (Milan), 122ª at Trenno and Ponte San Pietro and the Sezione Difesa SP.3 of Cascina Malpensa.

Gruppo Aviazione da Difesa Aerea Centrale

This unit was formed on 25 September 1916, at the command of *cap.* Adolfo Resio, based in Florence. In 1917 it was formed by the 105ª Squadriglia of Florence and the Sezioni Difesa of Ravenna, Rimini, Cairo Montenotte, Novi Ligure and Bologna, at the command of *magg.* Ettore De Carolis. In late 1917 it was based at Bologna. It was disbanded on 1 March 1918.

Gruppo Difesa Aerea Veneto-Emilia

Probably formed after the abolition of the preceding unit, the Gruppo Difesa had on charge the Sezioni Difesa of Bologna, Rimini and Ravenna, and since August 1918 it included also the 5ª Sezione Difesa of Padua. Its commander was *cap.* Umberto Venanzi.

Gruppo Aviazione da Difesa Meridionale

Also called Gruppo Squadriglie Difesa Aerea Italia Meridionale, this unit was based in Rome, and it was formed in September 1916, with commander *magg.* De Masellis, having on strength the Sezioni of Foligno, Jesi, Ravenna and Rimini. It also controlled the 103ª Squadriglia at Brindisi until this unit was assigned to the command of the Navy. In January 1918 its commander was *magg.* Eginardo Bongiovanni, and its units were the Sezioni of Ancona, Jesi and Terni, the 107ª Squadriglia of Roma and the 101ª of Bari. It was disbanded on 1 March 1918.

Groups of the Regia Marina
I Raggruppamento Idrovolanti

This unit, 1st Seaplanes Group, appeared as a nominal command in late 1916, under TV Giacomo Mancini. In April 1917, as it received its first FBA flights, it passed under the command of *cap.* Ivo Varanini, who remained until the unit was disbanded. Second in command was *ten.* Armando Ferrini. Reporting to the *Comando in Capo del Dipartimento Marittimo di Spezia*, (Command of the Spezia naval department) this unit had Sezioni FBA at San Remo, Porto Maurizio, Varazze, Spezia, Livorno, Piombino and Civitavecchia plus the Sezione Caproni at Campiglia, that later were upgraded to Squadriglie: in March 1918 the 266ª, 267ª and 268ª, in April the 269ª, in May also 271ª, 272ª, 273ª and 274ª, in July 285ª and 242ª da Difesa at Sarzana, and finally on the eve of armistice the Sezione Caproni at Campiglia became 202ª Squadriglia. The Raggruppamento

ITALIANI!

I vostri fratelli del Regno hanno riportato, sul Carso, una NUOVA GRANDE VITTORIA PER LA VOSTRA REDENZIONE.

Dinanzi all'impeto del sangue italiano, che pure scorre nelle vostre vene, le difese austriache sono crollate.

Più di 10.000 AUSTRIACI sono stati fatti PRIGIONIERI, fra i quali un generale comandante di brigata, un comandante di reggimento, e circa 350 ufficiali; 12 cannoni sono stati presi, insieme ad una grande quantità di munizioni, di materiali da guerra e di quadrupedi.

Questi sono i primi frutti conseguiti dagli italiani in meno di tre giorni, dal 1° al 3 del corrente Novembre.

Esultate! L'ITALIA, la vostra gran Madre, avanza. Si avvicina il termine delle vostre sofferenze.

4 novembre 1916.

Above: A leaflet in Italian, dropped on 4 November 1916 over Trieste, claiming victory in the 3rd Battle of the Isonzo, claiming 10,000 Austrian prisoners, which is true, but not reporting the Italian losses of 67,000 casualties, including 11,000 killed. (Archive Longhi)

progressed from 157 reconnaissance flights in April 1917 to 1,188 reconnaissance and convoy escort flights in July 1918. It was disbanded in May 1919, replaced by the constitution on 1 June of the *Comando di Aeronautica Alto Tirreno* of the Regia Marina.

II Raggruppamento Idrovolanti

Based at Naples, the II Raggruppamento Idro, 2nd Seaplanes Group, at the end of 1917 was at the command of *cap.* Ferruccio Capuzzo, replaced in October by *cap.* Alfredo De Luce. It had the seaplane flights of Naples, Ponza and Sapri and reported to the *Comando in Capo del Dipartimento Marittimo di Napoli* (Command of the Naples naval department). In 1918 this unit co-ordinated Squadriglie 275ª at Ponza, 276ª at Naples and 277ª at Sapri, and the Sezione Idrocaccia (seaplane fighter flight) at Lucrino.

III Raggruppamento Idrovolanti

The III Raggruppamento Idro, assigned to the *Comando dei Servizi per la Regia Marina in Sicilia*, was established in October 1917 based at Messina, its commander at the end of 1917 was *cap.* Vincenzo D'Ippolito and its units were the Sezioni FBA of Milazzo, Palermo, Syracuse, Catania, Giardini and Taormina, that progressively in early

1918 became Squadriglie. In the summer of 1918, the Raggruppamento included Squadriglie 270ª at Palermo, 280ª at Milazzo, 281ª at Giardini, 282ª at Catania, 283ª at Syracuse, 284ª at Trapani. *Cap.* D'Ippolito was killed in a flying accident on 10 August 1918 at Syracuse. *Cap.* Umberto Martellucci replaced him.

Gruppo "Aldo Giannelli" – 7° Gruppo Bombardamento

This naval group was in formation at the end of the war, at San Vito dei Normanni with squadrons of Caproni Ca.5 that were to form a large strategic force to strike the Austrian fleet in port. It was named after a fallen naval aviator. After the departure of American personnel, it remained at Brindisi at the command of STV Pietro Sandrelli, then, in May 1919, TV Tommaso Surdi and with just ten pilots, all of the Regia Marina, among them STV Giovanni L'Abate, Giuseppe Marini, Agostino Scotto di Marco, *2° nocc.* Carlo Incandela, *fuochisti* Ottavio Saroldi and Primo Zucca and a couple of Caproni 5 formerly of the US Navy. It never went into action.

Scuole di volo

The flying schools also were organized with a system of Gruppi, groups, which in 1917 included:

I Gruppo: Mirafiori, Venaria Reale
II Gruppo: Busto Arsizio, Cameri, Cascina Costa, Cascina Malpensa
III Gruppo: Coltano, San Giusto, Centocelle
IV Gruppo: Foggia Sud, Foggia Nord, Gioia del Colle
Gruppo idrovolanti: Sesto Calende, Passignano, Orbetello.

Above & Right: Pilot sleeve insignias.

Part 4: The Squadriglie up to 15 April 1916

Squadriglie da ricognizione e combattimento (Reconnaissance & Combat Squadrons)

1ª Squadriglia

Above: A Blériot of the 1ª Squadriglia showing the first national marking, a black circle. The first pilot from the left is ten. Ettore Croce. (Archive Caliaro)

Formed in 1912, the 1ª Squadriglia Blériot pre-war was based at Mirafiori, then at Venaria Reale, both near Turin, and in March 1915 it moved to Aviano. When Italy entered the Great War, 24 May 1915, it was part of the I Gruppo and moved to Chiasiellis, in the plain between Palmanova and Udine, assigned to the 3ª Armata. With commander *cap.* Costantino Quaglia were pilots *cap.* Filippo Sisto, *cap.* Mario Stanzani, *s.ten.* Alberto Verona, *maestro d'arme* Vinicio Muraro, with Blériots 240, 249, 251 and 300.

On 24 May 1915 *cap.* Quaglia flew the first mission, dropping bombs over Sagrado. In June the unit flew many reconnaissance missions carrying as observers *ten.* Enrico Giorgetti and *cap.* Ferdinando Sanità. On 4 June *cap.* Quaglia made another raid dropping bombs on Mount San Michele. On 12 June the squadron got Blériot 303, that replaced another aircraft crashed on landing by Muraro on 3 June. The obsolescence of the Blériot made other actions impossible, and the 1ª Squadriglia was disbanded on 25 June 1915.

A new 1ª Squadriglia was formed in July at Pordenone equipped with Farman 1914 with Renault 80 hp engines 515, 518, 519, 521, 522, 523 at the command of *cap.* Carlo Melegari, then *cap.* Amedeo Ferraro. The new unit went to war as part of the III Gruppo with pilots *cap.* Ferraro, *s.ten.* Enrico Cottino and Orlando Ferroni, *serg.m.* Filippo Palange and *soldato* Angelo Alberto, and the observer *cap.* Armando Giammarco.

On 24 July 1915 Farman 522 was lost in a landing accident. On 7 August the 1ª Squadriglia moved from Pordenone to Aviano, then on 23 August to S. Maria la Longa, assigned to II Gruppo, replacing the 2ª Farman that

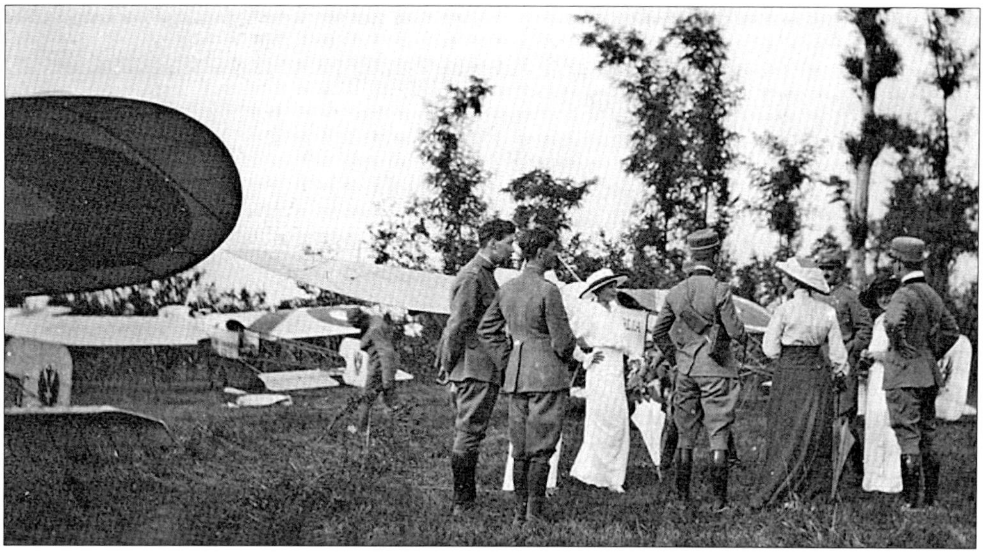

Right: Young ladies visiting the 1ª Squadriglia in the first days of the war. On the rudder the Blériots show the national insignia, an eagle with the shield of Savoy, introduced in April 1913.

was assigned to the Comando Supremo for the defence of Udine, then back to the III Gruppo on 25 August. That day the unit flew its first war mission, a flight over Aisovizza by *cap.* Ferraro. However on 25 August the Austrians raided Brescia and the 1ª Squadriglia was rushed to the defense of that city, the center of Italian light weapons production.

The role of air defense then passed to the 3ª Squadriglia Aviatik, so on 6 October 1ª Squadriglia went to Verona, leaving in Brescia a flight of two Farmans, 521 and 524, losing along the way Farman 518 of *s.ten.* Cottino, crashed at Mantua. On 7 October the unit flew its first mission on the new front of the 1ª Armata. On 11 November some of its planes bombed Volano.

On 1 December 1915, like all similar units, it got the new name 1ª Squadriglia da Ricognizione e Combattimento (reconnaissance and combat) and went back to the reborn III Gruppo. It got new equipment, Farmans 757, 759, 803, 818, 819 and 844. On 19 December *cap.* Ferraro left command to *cap.* Ferdinando Bonazzi. On 20 and 21 December the unit bombed barracks near Forte Por, starting a series of raids that

Above: Blériot 300, probably used for training after service with the 1ª Squadriglia, now with the national colors of green, white, and red on the rudder. (Archive Schifano)

Above: Farman 518 and 521 with Renault engines of the 1ª Squadriglia.

made it become one of the crack units of Italian aviation. During 1915 the 1ª Farman flew 151 war missions and *s.ten.* Orlando Ferroni was awarded a *Medaglia di Bronzo* (bronze medal).

On 1 January 1916 the squadron had on strength pilots *cap.* Bonazzi, *s.ten.* Ferroni, *caporali* Francesco Bobbio and Umberto Ferro, *soldato* Michele Riello, and observers *tenenti* Paolo Avogadro and Antonio Benedetti, *sottotenenti* Mario Poli, Guido Roppa Sassoli and P. Rossetti.

On 10 January all five planes flew a massed sortie, they overflew Trento, three aircraft bombed the Austrian airfield at Gardolo while another bombed Rovereto railroad station. On 10 February the unit bombed enemy encampments at Calliano. On 13 February a Farman suffered engine failure, but it glided back over the lines and crash landed with no injuries to its crew. On 20 February Farman 757 was lost in an accident, while Farmans 911 and 1473 arrived. On 21 February *cap.* Bonazzi and *ten.* Poli had a brief combat with an enemy airplane. On 27 March five Farmans with pilots *cap.* Bonazzi, *s.ten.* Cottino, *serg.* Bobbio, *cpr.* Riello and *s.ten.* Ferroni took off on alarm to intercept a raid on Verona, and they managed to intercept Austrian airplanes, that got away after a prolonged exchange of fire. On the way back, they met two more and tried to attack them with machine-gun fire, but they failed to stop them from dropping bombs over Verona. The Farman was much inferior to the Lloyds and Brandenburgs in service with the Luftfahrtruppe.

On 1 April 1916 enemy airplanes attacked the Farmans of *cap.* Bonazzi and *cpr.* Riello over the plateau of Lavarone, but the Italians got away after having fired many shots, still *cap.* Bonazzi managed to complete his photo reconnaissance. On 10 April two Farmans were tasked with a reconnaissance to Trento, *ten.* Cottino with observer *ten.* Carlo Emanuele Parini (an *irredento*, that is to say an Italian with Austrian nationality, who used the fictitious name Carlo Emanuele Aprato) had a combat with two enemy aircraft above Mount Finocchio, one of the Austrians left the fight, possibly hit, while *serg.* Bobbio with observer *ten.* Luigi Sella met an enemy plane flying to Verona, had two combats over Verona, and pursued the enemy that got away in the Adige valley.

On 15 April 1916 the system of designation of Italian Squadriglie was changed and the 1ª Farman became 31ª Squadriglia.

2ª Squadriglia

Above: Blériot 286 of the 2° Squadriglia showing the national markings in use in the first weeks of the war.

The 2ª Squadriglia Blériot, was based at Venaria Reale and in January 1915 it moved to Aviano.

It was part of the I Gruppo, at the outbreak of war its commander was *cap. pil.* Ercole Capuzzo with pilots *cap.* Ferdinando Sanità, *ten.* Luigi Reggio, *serg.m.* Luigi Bertello and Giuseppe Screpanti and observer *ten.* Enrico Giorgetti. It was equipped with Blériots 32, 33, 253, 282 and 286, based at Chiasiellis, operating for the 6th Army Corps.

On 2 June 1915 it was assigned to the II Gruppo. Active in reconnaissance missions south of Palmanova and in bombing raids against the Redipuglia power station and Ronchi and Monfalcone railroad stations, it also distinguished itself taking many photographs. On 7 June *cap.* Sanità dropped four bombs on Gorizia, but ground fire hit his airplane. On 12 June *serg.m.* Bertello flew Blériot 32 to "*test the plane with the new markings*" that is, black circles and diagonal bands under the wings.

On 18 June Capuzzo left command to *cap.* Gaspare Bolla, a famous Cavalry officer. On 23 June, supporting the Italian offensive, the unit flew five bombing missions by Sanità and Bertello (two each) and Reggio. On 2 July a Blériot bombed Duino. Three days later *ten.* Reggio and Giorgetti took off for a mission in the rain but they had to give up, because after 32 minutes their Blériot had hardly reached 2,000 feet. It was evident that the Blériots were too old for modern warfare. On 8 July *cap.* Novellis, commander of the I Gruppo, wrecked Blériot 253. On 16 July another pilot arrived, *s.ten.* Umberto Re. On 18 July the whole squadron moved to Aiello for a special mission, that was cancelled on account of bad weather. In the flight back to base *cap.* Gaspare Bolla crashed to his death in Blériot 32. This was the first casualty of Italian heavier-than-air aviation at the front.

Commander of the 2ª Squadriglia became *cap.* Sanità. On 27 *serg.m.* Bertello attacked with bombs a kite balloon, with no apparent result. On 24 July the order came to disband the 2ª Squadriglia, and this took place on 31 July 1915, when the unit had Blériots 286, 295, 299, and 302. It had flown 18 bombing and 23 reconnaissance missions; *cap.* Bolla was awarded the *Medaglia d'Argento* (silver medal).

Meanwhile on 18 July a new 14ª Squadriglia Farman was formed at Pordenone, with commander *cap.* Luigi Chiappelli, this unit in August took up the now unused number 2 becoming 2ª Squadriglia Farman. Mobilized on 6 August 1915, it received a new commander on August 10, *cap.* Michele Masi, pilots were *cap.* Chiappelli, *s.ten.* Sestino Mauro, *cpr.* Franz Almerigi, *serg.* Costantino Rossi, *soldato* Vittorio De Lucchi, with observer *cap.* Achille Sergardi. It was equipped with Farmans 425, 430, 431 and 441 with Fiat 100 hp motor.

The 2ª Farman went to the I Gruppo at Chiasiellis for just twelve days, then it went to the II Gruppo at Campoformido since 19 August, but on 23 August it was assigned to the

Above: A bad landing for Blériot 295 of the 2ª Squadriglia.

Above: Blériot 293 also had a bad landing.

Above: Farman MF 430 is now carrying a machine gun. Its engine is a 100 hp Fiat.

Comando Supremo for the defence of Udine, reinforced with two Nieuport X biplanes, coming from the 8ª Squadriglia with pilots *capitani* Guido Tacchini and Matteo Scarpis and *ten.* Francesco Baracca. It then became part of the *Gruppo Difesa di Udine* of the Comando Supremo. On 1 September the Sezione Nieuport gave *cap.* Scarpis and received *cap.* Oreste Salomone and *ten.* Domenico Bolognesi.

On 19 November the 2ª Squadriglia put up five Farmans and four Nieuports against a raid by five Austrian airplanes against Udine. On 1 December 1915, like all similar units, it got the designation 2ª Squadriglia da Ricognizione e Combattimento. On 15 December the Sezione Nieuport left the unit. During 1915 it flew 279 war flights and had 7 air combats. The *Medaglia di Bronzo* decorated *cap.* Chiappelli and *serg.* Screpanti.

On 1 January 1916 the 2ª Squadriglia had pilots *cap.* Masi, *s.ten.* Mauro, *serg.m.* Costantino Rossi, *caporali* Franz Almerigi, Antonio Lovadina, Tersilio Taliani and Elio Vendruscolo, and observers *cap.* Achille Sergardi, *tenenti* Ercole Messi, Michele Martinelli and Luigi Coniglio. In February it had Farmans 425, 430, 433, 439, 526, 533, 561, 809, 849 and 850. On 7 April one of its Farmans took part in the air combat above Udine during which two enemy Brandenburgs were shot down, but the Farman, possibly hit by enemy fire, crashed killing *ten.* Martinelli, while *cpr.* Taliani was wounded. The *Medaglia d'Argento* was assigned to the two airmen. The 2ª Squadriglia Farman became 33ª Squadriglia on 15 April 1916.

Farman 439: in its original version the Farman MF had no cloth cover on the fins.

Below: The Farman that crashed on 7 April 1916, killing *ten*. Martinelli, after combat with Austrian Brandenburgs.

3ª Squadriglia

Above: Blériot 242 of the 3ª Squadriglia carrying two small bombs under the fuselage.

The 3ª Squadriglia Blériot was based at Cuneo pre-war and in January 1915 it moved to Aviano as part of the I Gruppo. On the eve of war, it had commander *cap.* Egidio Carta and pilots *cap.* Luigi Falchi, *ten.* Umberto Piazzi, *serg.m.* Enrico Mazzetti and Vincenzo Arone, with Blériots 2, 34, 242, 283, 298, and 304, moving on 21 May to San Vito al Tagliamento.

It flew its first mission, sending three Blériots on reconnaissance, at dawn on the first day of war, 24 May 1915. On 25 May Mazzetti dropped two bombs, and his Blériot was hit by enemy ground fire. For the first time, Falchi dropped leaflets urging the population not to co-operate with the Austrians (this kind of propaganda was stopped, when it turned out to be counter-productive). Two days later Falchi was forced down in Blériot 242 "Roma" because a bullet hit a fuel pipe but the pilot managed to fix the damage and return to base. On 29 May the squadron moved to a field between Mortegliano and Chiasiellis, flying a relevant number of missions in the area between Gorizia and Monfalcone reporting there a minimal amount of Austrian defenses.

On 13 June, Carta, Piazzi and Mazzetti flew a bombing mission against enemy batteries, each Blériot dropping four bombs, Mazzetti with Blériot 299 suffered enemy ground fire which wounded him and put his engine out of action, but the pilot managed to glide and land within the Italian lines. Mazzetti received the *Medaglia d'Argento "motu proprio"* (that is, directly, without passing through commands) from king Victor Emmanuel III who had seen the event: it was the first medal for valor to an airman in the war. The airplane was put back in service. On 17 June Piazzi bombed Doberdò, with two bombs aimed at the main square of the village. On 18 June, as the 1ª and 14ª Squadriglia Blériot were disbanded, the 3ª receiving their airplanes.

On 23 June Carta, Piazzi and Falchi flew three bombing missions, against enemy trenches and Falchi a fourth one against a ship in the harbor of Trieste. On 27 June *cap.* Falchi with Blériot 242 was hit over Doberdò in a bombing raid, with a pierced fuel tank he landed at S. Maria la Longa. On 28 June the 3ª Blériot moved to Quota 43 (Hill 43) to support the 1st Battle of the Isonzo. The Squadriglia was disbanded on 16 July 1915, when it had flown 38 war missions, the last ones on July 5, when two airplanes bombed Oppachiasella and a reconnaissance on 16 July.

On 26 September 1915 formation began at Brescia of a new 3ª Squadriglia Aviatik, which on 1 December became the 3ª Squadriglia Caccia.

Left: A Blériot of the 3ª Squadriglia has smashed its landing gear. On the rudder there is the black eagle, and the black circle and bar on the wing confirm that this is a wartime photo.

Above: Soldiers loading Blériot 34 of the 3ª Squadriglia with small bombs.

Above: Blériot 283 of the 3ª Squadriglia, on the rudder the national colors have been painted, but the black eagle has been retained.

4ª Squadriglia

Above: The Blériots of the 4ª Squadriglia at Bazzera, supposedly protecting Venice. (Archive Apostolo)

The 4ª Squadriglia Blériot, of the II Gruppo, at the outbreak of war moved from Padua to Bazzera, close to Venice, reporting to the Command of the Maritime Station of Venice. With commander *cap.* Guido Olivo there were pilots *tenenti* Cesare Buzzi, Umberto Clerici, Eraldo Parenti and *caporale* Ferruccio Rossi, and equipment consisted of Blériots 80 hp 5, 15, 21, 22, 36 and 281. All through the spring and summer this unit flew little, doing at best one or two test flights each day, while one after the other *cap.* Olivo, *ten.* Parenti and Buzzi were detached to the Malpensa flying school, so the Squadriglia was left with just two pilots. It flew a total of 103 missions, just reconnaissance over the sea around Venice.

The 4ª Squadriglia was disbanded on 8 September, when the defense of Venice was assigned to the French Nieuports. It was the last Blériot squadron.

A new 4ª Squadriglia Ricognizione and Combattimento Farman was formed at La Comina airfield on 18 January 1916 for the II Gruppo. It was activated on 20 January 1916 and deployed at Cavazzo Carnico, that it reached on 10 February, being the first, and for a long time, the only Squadriglia operating in the northern Carnia region. Its force consisted of commander *cap.* Umberto Venanzi with pilots *s.ten.* Angelo Tucci, *serg.m.* Gualberto Giuntoli, *sergenti* Franz Almerigi, Renato Ghizzoni and Antonio Piccin, *sold.* Renato Giampietro, and observers ten. Federico Casabella, Luigi Coniglio and Federico Durini. It was equipped with Farmans 816, 824, 826, 827, 828, 829, and 901. It flew its first war mission on 21 February 1916.

It didn't fly on account of bad weather until 18 March 1916. *Cap.* Achille Sergardi became its temporary commander, but Venanzi returned by the end of the month. The 4ª flew frequent reconnaissance missions and even interception flights.

The 4ª Squadriglia became 29ª Squadriglia on 15 April 1916.

A Farman of the 4ª Squadriglia with Antonio Locatelli, a young pilot who became one of Italy's most important aviators.

Above: Personnel of the 4ª Squadriglia with Farman 901, in the background the harsh mountains of the Carnia region. (Archive Caliaro)

Right: The elegant eagle of the first national insignia of Italy.

Below: An early production Farman model 1914, or Farman MF, with the first type of national insignia, an eagle with the shield of Savoy on the rudders.

5ª Squadriglia

Above: *Cap.* Piccio, standing, with a Nieuport Monoplane of the 5ª Squadriglia, the war marking of bar and circle are painted white on the darker wings of the Nieuport. The man in the cockpit is *ten.* Ercole Messi.

The 5ª Squadriglia Nieuport monoplanes, based at Busto Arsizio, moved to Pordenone in January 1915. Assigned to the III Gruppo, at the outbreak of war it was based at Campoformido, attached to the Comando Supremo. It was an élite unit (in 1913 Baracca flew there, in 1914 it won a competition among Squadriglie). In May 1915 its force included commander *cap. pilota* Pier Ruggero Piccio, *cap.* Luigi Tagliasacchi and Dario Ungania, *tenente* Giulio Palma di Cesnola, flying Nieuports 80 hp 116, 118, 122, 126 and 260 and Nieuport 100 hp 93.

On 25 May *cap.* Piccio and Tagliasacchi bombed the factories of Monfalcone. Then the unit bombed the electric power station of Nabresina, as it was feared that it was used to electrify the barbed wire lines of the Austrians. The 5ª Squadriglia was a very active unit: on 5 June alone, it flew ten war missions, three with Ungania, four with Palma and three with Piccio. It carried as observers *tenenti* Carlo Berliri Zoppi, Ercole Messi and Giuseppe Rossi and *cap.* Gallina. It originally used as bombs 90-mm artillery shells, then a way was found to also drop the big 162-mm shells through a hole in the Nieuport fuselage floor.

On 11 June the 5ª Squadriglia moved from Campoformido to Santa Maria la Longa, and kept on its air offensive against Nabresina. Piccio and Messi flew three bombing missions on 23 June, and more the following days. On 27 June Piccio and Berliri using a pistol fought against an enemy airplane carrying a machine gun. On 8 July Piccio and Messi, active as usual, flew the first night reconnaissance mission. The Nieuport monoplane, however, was obsolete and the 5ª Squadriglia was disbanded on 31 July 1915. In 68 days at the front, it had flown 112 war missions. Its disbandment was a plus, because its élite pilots were promoted to better aircraft, assigned to the 3ª Squadriglia da Offesa, with the new Caproni bombers. Its *palmarès* included Silver and Bronze medal to Messi, Bronze to Palma di Cesnola and Ungania.

A new Voisin Squadriglia, 16ª, was formed at Aviano on 9 August 1915 and it took up the number 5 that was now available. Its commander was *cap.* Leopoldo De Rada, a veteran of the Libyan war of 1911, pilots *ten.* Alessandro Raffaelli, *cap.* Edoardo Carignani di Valloria, *serg.m.* Michele Ballerini and *m.llo* Gaspare Pollini, its only observer was *ten.* Fausto Pesci, soon joined by *cap.* Enrico Giorgetti and *ten.* Enrico Calma.

The new 5ª Squadriglia on 13 August went to Campoformido, in the II Gruppo, where it replaced the disbanded 7ª Nieuport, with four Voisins with Salmson Canton Unné engine: 377, 379, 380 and 381.

Right: Voisin 378 in marked "XVI^a Squadriglia" but when this unit became operational it was named 5^a Squadriglia.

On 15 August 1915 this unit flew its first reconnaissance over Sesana, the next day its first bombing mission. On 19 August it was temporarily assigned to the I Gruppo, (this assignment becoming final on 12 September) moving to Chiasiellis airfield attached to the 3ª Armata. It flew armed reconnaissance missions dropping its bombs on targets of opportunity, facing the fire of Austrian artillery, but also of the Italian one, still deficient in aircraft recognition. On 26 August Carignani and Giorgetti met a biplane which they identified as a Nieuport (there were no Nieuport biplanes yet on the Italian front) which fired at them repeatedly. On 14 September *cap.* De Rada, having as observer *col.* Riccardo Moizo, had a duel with an enemy plane over Aquileia. On 29 September over the Isonzo river, five Austrian airplanes attacked the Voisin of Carignani and Giorgetti, who believed that one enemy was hit, quitting the combat. By the end of September, 5ª Squadriglia was the leader in terms of bombs dropped and photographs taken.

On 7 October with its three available aircraft, it went with the other Squadriglie of the I Gruppo to bomb the command of Archduke Joseph August at Kostanjevica. Due to bad weather only the aircraft of Carignani with observer *cap.* Giorgetti reached the target and dropped four 112-mm bombs, also taking 12 photos. Then an enemy airplane approached it and a duel began, the Italians fired 64 rounds and drove off the attacker. Two aircraft repeated this action the next day with the crews of *cap.* De Rada and *ten.* Pesci, *cap.* Carignani and *ten.* Calma.

One more air combat took place on 12 October, *serg.m.* Ballerini and *cap.* Pesci fired 101 rounds and returned to base with a damaged aircraft.

On 1 November 1915 Voisin 377 with pilot *cap.* De Rada and *cap.* Calma coming back from a mission on Gris had an engine failure and force landed, the two airmen were lightly

Right: Voisin 380 of the 5^a Squadriglia, unarmed but with a tube for dropping bombs, a device that disappeared quickly. (Archive Salvo Di Marco)

Above: Voisin 381 of the 5ª Squadriglia, bombs are laying on the ground. (Archive Caliaro)

Above: Pilots of the 5ª squadriglia; behind them a hangar under construction.

Above: The Voisin of *serg*. Ranucci and *ten*. Lanza di Trabia, crash landed near Venice on 16 February 1916 after a mission over Trieste.

injured. On 1 December 1915 the unit took the new name 5ª Squadriglia da Ricognizione e Combattimento. At the end of 1915 it received Voisins 980, 1026, 1039 and 1098 with Isotta Fraschini engine. During 1915 the unit had flown 101 war missions.

On 1 January 1916 the 5ª Squadriglia had on strength pilots *cap.* De Rada, *cap.* Mario Stanzani, *cap.* Carignani, *aspirante* Ballerini, and observers *capitani* Pesci, Armando Fabretti and Calma.

The first flight of the New Year took place on 10 January, by the crew of Ballerini and Pesci, and enemy ground fire hit their Voisin. On 20 January Pesci with pilot *serg.* Vittorio Aquilino had a combat with an enemy monoplane that hit their Voisin with several bullets. On 16 February 1916 *serg.* Giovanni Ranucci and *ten. oss.* Ignazio Lanza di Trabia having flown for about one hour and a half, first in darkness and then in fog, going to bomb Trieste, discovered that their gas was running out. In order not to fall in enemy's land they decided to take a shortcut crossing the sea at low level in the fog, but they fell into the Adriatic. Lanza, a Sicilian Prince, waiting for help, while hanging to the hulk of his Voisin, wrote a letter to his mother expressing his joy to give his life for the Country.

On 1 April an enemy "Albatros" (probably a Brandenburg) attacked the Voisin of pilot *m.llo* Vinicio Muraro with gunner *cpr.* Giuseppe Ferrari, at the union of the Torre and Natisone rivers. During the combat the Italians fired 220 rounds. The enemy aircraft got away in gliding flight toward Gorizia, and official sources recognized this combat as an Italian air victory, although 5ª Squadriglia records did not claim it as such: "*Aspirante Muraro Vinicio and gunner caporale Ferrari during a cruise engaged combat with an enemy Albatros. 200 machine gun rounds were fired at a distance of 200–400 metres. The enemy airplane was seen going down out of control behind Mount San Michele.*"

On 6 April 1916 a Fokker attacked the Voisin of *serg.* Ermete Mongardi and *ten.* Alfredo De Luce above Merna, hitting it in the struts, but the Italians defended themselves and drove it off. The 5ª Squadriglia became 25ª Voisin on 15 April 1916. Operating over the Carso plateau, where action was harshest, and often faced by enemy aviation, this unit proved its exceptional pugnacity and many medals were awarded to its airmen: *Medaglia d'Argento* for *capitani* Enrico Calma, Edoardo Carignani and Pesci, *tenenti* Fabretti and Lanza di Trabia, *serg.* Ranucci; *Medaglia di Bronzo* for *cap.* Leopoldo De Rada.

6ª Squadriglia

Above: Nieuport Monoplane 133 of the 6ª Squadriglia. (Archive Achille Vigna)

The 6ª Squadriglia Nieuport, of the II Gruppo, before Italy's entry in the war had commander *cap.* Oronzo Andriani, *tenenti* Francesco Baracca, Salvatore Calori and Raul Lampugnani, and *marescialli* Arturo Guglielmi and Luigi Rocchi at Pordenone, where it had arrived in January 1915 coming from Taliedo. At the outbreak of war, the unit moved from La Comina, Pordenone's airfield, to Campoformido being assigned to the II Gruppo and the 6ª Squadriglia took on strength also *brigadiere* Annibale Comazzi. With Andriani there were pilots *cap.* Filippo De Rossi Di Santa Rosa and *m.llo* Rocchi and observers *cap.* Gabriele Vallo and *tenenti* Enrico Calma and Armando Giammarco, with Nieuport monoplanes 108, 111, 112, 259 and 291, while cap. Lampugnani was posted to train on the Macchi Parasol.

On 25 May 1915 all three pilots of the 6ª Squadriglia bombed the electric power station of Monfalcone, and in the following days they flew other reconnaissance and bombing missions. In the first days of June the squadron moved from Campoformido to Bicinicco and one week later, June 11, to nearby S. Maria la Longa for the 1st Battle of the Isonzo, at the same time passing to III Gruppo. It operated mainly in spotting for the artillery. On 20 June Nieuport 112 crashed due to an engine failure, its pilot Comazzi escaped unharmed, observer *ten.* Vittorio Giovine was wounded. On 23 June another engine failure crippled Nieuport 108 of *cap.* De Rossi di Santa Rosa engaged in a reconnaissance over Sagrado, and the airplane made a forced landing in a swamp.

On 6 July two Nieuports attacked the hangars of Aisovizza: they were *m.llo* Rocchi with *cap.* Giammarco and *brig.* Comazzi with *ten.* Berliri Zoppi, who dropped eight incendiary bombs. On 10 July the Squadriglia was assigned to the artillery command of the VII Corpo d'Armata.

On 27 July there was the first loss due to enemy action of the Italian air service. Austrian anti-aircraft cannons shot down, near Lake Pietrarossa, the Nieuport of *m.llo* Rocchi with *Cap. di Fregata* Riccardo Cipriani, a naval officer, killing them. On 6 August *cap.* Andriani with the observer *ten.* Giovine spotted for the 305-mm howitzers firing from Isola Morosini.

The 6ª Squadriglia Nieuport was disbanded on 27 August. During its short period of war action two Silver Medals were assigned to *CF* Cipriani and *m.llo* Rocchi, and three Bronzes to *cap.* Andriani, *cap.* Vallo and *ten.* Berliri Zoppi.

One month later, on 25 September 1915, a new 6ª Squadriglia was formed at Pordenone equipped with nine Farmans with Fiat 100 hp engine (only known serial is 856). It was assigned to the II Gruppo, joining the 9ª Squadriglia, and being based at Campoformido. On 7 October the 6ª Farman moved to Chiasiellis: its commander was *cap.* Guido Olivo with *s.ten.* Achille Romice, *caporali* Lorenzo Rossi (later a test pilot with Pomilio, killed in a test flight) and Roberto D'Anna and *soldati* Antonio Locatelli and Arturo Dell'Oro with six Farman Fiat. This unit flew its first war mission on 8 October: a reconnaissance to Podgora and three patrol flights. On 12 October it got its observers: *cap.* Enrico Giorgetti, *ten.* Mario Ajmone Cat (the future commander of the Italian air force in Ethiopia and Chief of Staff after WW2) and *sottotenenti* Umberto Livi, Giuseppe Costa and Lucifero Antonelli.

Above: Nieuports of the 6ª Squadriglia showing the war markings, a white bar and a roundel.

The first bombing mission took place on 18 October 1915, followed by many more in November, dropping bombs and steel darts. In one such mission, on 12 November, the airplane of *cpr.* Dell'Oro suffered an engine failure and force landed. On 18 November 1915 there was the first air combat: *m.llo* Nino Santerni and *ten. oss.* Livi at a height of 9,000 feet fought against an enemy airplane flying toward Udine, which they drove off over the Isonzo. Two days later, enemy ground fire hit the airplane of *cpr.* Rossi and *ten.* Ajmone Cat over Aisovizza. On 23 November *serg.* Dell'Oro and *ten.* Costa attacked an enemy airplane over the valley of the Vippacco, they had to stop because their machine gun jammed, then they went on for two hours in a difficult mission, during which they discovered a large enemy column. On 1 December the unit took the name 6ª Squadriglia da Ricognizione e Combattimento. During 1915 it flew a total of 69 war missions, with three air combats.

On 1 January 1916, still based at Chiasiellis, the 6ª had *cap.* Guido Olivo, commander, pilots *s.ten.* Nicolò Parodi, *m.llo* Santerni, *caporali* Dell'Oro, Locatelli, Lorenzo Rossi, and observers *cap.* Giorgetti and *tenenti* Ajmone Cat, Antonelli, Costa and Livi.

On 10 January 1916 during the first war flight of the year, a difficult reconnaissance over Mount Nero, enemy artillery hit the propeller and nacelle of the aircraft of *s.ten.* Parodi and *ten.* Ajmone Cat. On 11 January *serg.* Rossi and *cap.* Giorgetti had a long duel with an enemy aircraft, they expended all their ammunition and were hit several times. The 6ª Squadriglia was tasked also with the defense of Udine and the escorting of artillery aircraft.

On 29 January *s.ten.* Parodi and *ten.* Costa had a duel with an enemy airplane that got away gliding toward Aisovizza. Another combat took place on 6 February, when the Farman of *ten. oss.* Livi fought against two enemy aircraft, which were driven away over the Vodice after having repeatedly hit the Italian plane. On 14 February Locatelli and Ajmone Cat had a combat with an enemy airplane that they claimed as shot down. On 18 February Antonelli flew a difficult reconnaissance in the area of the Smocar and anti-aircraft guns hit his aircraft damaging it baldly. On 19 February *serg.* Rossi and *cap.* Giorgetti attacked once more an enemy aircraft at S.Lucia di Tolmino, while a Farman bombing enemy barracks near Slap was hit 34 times.

On 20 March 1916 the 6ª Farman was left as the only unit of the II Gruppo, because the 4ª moved to the separate sector of the Carnia. The 6ª Farman became 30ª Squadriglia on 15 April 1916. The *Medaglia d'Argento* was awarded to *cap.* Giorgetti, *ten.* Lucifero Antonelli, *ten.* Ajmone Cat, *ten.* Giuseppe Costa, *serg.* Lorenzo Rossi, *ten.* Alberto Livi, *s.ten.* Antonio Locatelli, *sold.* Dell'Oro and *m.llo* Nino Santerni.

Left: *Ten*. Calori in front of a Nieuport, in the cockpit is *cap*. Pier Ruggero Piccio, future ace and commander of the Italian Air Force.

Below: Nieuport monoplane 111 of the 6ª Squadriglia showing the black bar national marking on its wing. (Archive Giorgio Catellani)

7ª Squadriglia

Above: Nieuport Monoplane 292 of the 7ª Squadriglia.

The 7ª Squadriglia Nieuport, of the II Gruppo, at Pordenone, moved to Campoformido on 20 May 1915. Its commander was *cap.* Armando Armani, its pilots were *cap.* Domenico Mondelli (a black man from Abyssinia, called "*il negro meraviglioso*" for his outstanding courage, later a General), *ten.* Umberto Venanzi, *serg.m.* Giuseppe Barattini, with *s.ten. osservatore* Luigi Coniglio. It was equipped with Nieuport monoplanes 125, 223, 226, 227, 254 and 292.

During the move of 20 May, the airplane of *serg.m.* Barattini crashed. On 25 May the 7ª Squadriglia began its war service with a reconnaissance by Venanzi over Salzano. On 2 June Mondelli in reconnaissance to Greifenberg had to force land at Corno di Rosazzo after enemy shrapnel hit his engine. The unit's next war mission was on 5 June, when Venanzi with Nieuport 292 bombed Probacina. On the 7th Mondelli with Nieuport 254 bombed Britof. War missions of this unit were scarce: on 10 June Mondelli flew a reconnaissance to Gradisca, on the 13th Venanzi dropped bombs and steel darts on enemy encampments east of Gorizia. *Serg.* Barattini then returned to service ferrying Nieuport 119, and *ten. oss.* Michele Martinelli arrived.

On 2 July rifle fire hit the airplane of Venanzi and Martinelli: not Austrian, but Italian soldiers were the marksmen! Following actions were: 12 July bombing of enemy tents at Gorizia; 15 July an attack near the Gorizia graveyard; 17 July steel darts dropped on enemy tent camps at Ranziano and bombing of a kite balloon. In total from the declaration of war, 24 May, to 31 July 1915 this unit flew 27 missions, but higher up somebody criticized its commander who, strictly adhering to regulations, didn't fly any war mission, while the commanders of the other Squadriglie of the II Gruppo, 8ª and 9ª, flew to give an example to their pilots. On 4 August *ten.* Michele Martinelli had a duel with an enemy airplane. Due to the absolute obsolescence of its aircraft, the 7ª Squadriglia was disbanded on 17 August 1915.

A new 7ª Squadriglia (that initially was called 17ª) equipped with Voisins with Salmson Canton Unné 140 hp engines was formed at Aviano in early August and, after due training, it was activated on 28 August, joining the I Gruppo moving to S. Maria la Longa. Its commander was *cap.* Ercole Capuzzo, pilots *capitani* Enrico Franceschini, Ernesto Jacometti and Ermanno Beltramo, *m.llo* Gaspare Pollini. Assigned to the XIII Corpo d'Armata of the 3ª Armata, it had Voisins 585, 586, 587, 588, 589, 590.

It started its war service on 2 September with a reconnaissance from Sagrado to Monfalcone and a cruise over Monfalcone and mount San Michele. In mid-September this unit was placed under the direct command of the 3ª Armata and then it flew many reconnaissance and artillery spotting missions.

On 7 October its three serviceable aircraft took part with the other units of the I Gruppo in the bombing at Castagnevizza (Kostanjevica) against the command of Archduke Joseph August. Due to bad weather none of them reached their target so five aircraft repeated the action the next day. On 12 October two crews, *cap.* Jacometti and *cap.* Della Noce, *m.llo* Santerni and *s.ten.* Domenico D'Alessandro, sustained the first air combats and managed to drive the enemy away. From 17 October to 6 November the 7ª also had in force *cap. pil.* Quaglia. It received new aircraft, Farman 544 and 547 and Voisins with Isotta Fraschini 150 hp engine 1027, 1029 and 1083, and by December it had withdrawn its old Voisin Salmson 140 hp. Meanwhile V588 had crashed on 29 November, piloted by *cap.* Jacometti. On 7 November the aircraft of *cap.* Franceschini and *ten. oss.* De Bratti had two air combats in a single morning, both of them with no result. On 1 December 1915 the unit was named 7ª Squadriglia da Ricognizione e Combattimento.

The 7ª Squadriglia flew reconnaissance in January 1916 all the way to Trieste. On 9 January it received pilots *sottotenenti* Giovanni Righi, Guido Masiero, Armando Deidier and *serg.* Giuseppe Buffa. On 31 January 1916 *cap. pil.* Franceschini was wounded in the head by anti-aircraft fire, and a *Medaglia d'Argento* was assigned for this fact, his observer baron *ten.* de Bratti, an irredento (that is, an Italian from an Austrian territory), agreed to fly on the mission to its end, earning a *Medaglia d'Argento*.

On 16 February a Voisin was lost: it was the one of *cap.* Jacometti and *ten. oss.* Marc'Aurelio Ripamonti that failed to return from a bombing raid to Castagnevizza and force landed in enemy land, the two airmen being captured. (In many texts Ripamonti is wrongly included among the casualties of the Caproni raid of 18 February). On 28 February the 7ª Squadriglia received *cpr.* Flavio Torello Baracchini and *serg.* Paolo Nicosia. Baracchini made his first flight on 14 March for "machine gun test" with *magg.* Abiel Bethel Revelli di Beaumont, designer of the Fiat Mod. 1914 machine gun and of the Villar Perosa machine pistol.

On 4 March *serg.* Righi with observer *s.ten.* D'Alessandro had a combat with a Fokker over Reifenberg. The Austrian fighter hit the Voisin before flying away. Another combat took place on 31 March 1916, involving *cpr.* Baracchini and *ten.* de Bratti, who fired five magazines while their plane was hit. On 10 March the 7ª received pilot *s.ten.* Michele Ballerini who remained with this unit until 1 June 1916. Baracchini and *ten.* Guido Pacinotti suffered an engine failure on 9 April, but they managed to glide back to Lucinico, in Italian land, where enemy artillery smashed their grounded Voisin.

Meanwhile, on 23 March 1916 the 7ª Squadriglia got a new commander, *cap. oss.* Ruggero Ubertalli. When the units were re-numbered, it became 26ª Squadriglia. On 15 April 1916, it had in force pilots *s.tenenti* Ballerini, Righi, Masiero, *aspiranti* Deidier and Baracchini, *sergenti* Buffa, Bianchi, Nicosia, observers *cap.* Ubertalli, commander, *magg.* Della Noce, *cap.* de Bratti, *tenenti* D'Alessandro, D'Ippolito and Pacinotti. The palmarès of the 7ª Voisin included *Medaglia d'Argento* to *cap.* Franceschini and Capuzzo, *ten*. de Bratti, *Bronzo* to *cap. oss.* Carlo Berliri, *cap. oss.* Ubertalli, *cap.* Jacometti, *ten.* Ripamonti and *ten.* D'Ippolito.

Above: Domenico Mondelli with a Nieuport Monoplane. After being transferred to infantry, in August 1917 he was the only Italian officer to enter Castagnevizza during the 11th Battle of the Isonzo.

Above: Voisin 589 of the 7^a Squadriglia being prepared for a mission, soldiers are attaching fins to a 162-mm bomb. (Archive Caliaro)

The same Voisin 589 ready to fly; a Fiat 1914 machine gun is fitted on the front tripod. (Archive Caliaro)

Above: Voisin 590 and 585 of the 7ª Squadriglia.

Above: Voisin 608, here still in the field of its manufacturer, SIT in Turin, was assigned to the 7ª Squadriglia, on its fuselage the construction number, 143, is marked with chalk.

8ª Squadriglia

Above: Nieuport 10 387 of the 8ª Squadriglia armed with a Fiat gun.

The 8ª Squadriglia Nieuport based at Padua moved in January 1915 to Pordenone. At the outbreak of war, it was part of the II Gruppo at Campoformido. Its commander was *cap.* Adolfo Resio, its pilots *cap.* Attilio Calderara (later transferred to airships), *ten.* Antonio Bosio (future Chief of Staff of the Regia Aeronautica), *ten.* Gaetano Coniglio, *brig.* Annibale Comazzi, observer *ten.* Andrea de Bratti, with Nieuports 133, 222, 257. This unit flew three reconnaissance missions on the first day of war, 24 May 1915. It soon received *ten.* Marc'Aurelio Ripamonti. On 9 July de Bratti, on a Nieuport piloted by *cap.* Resio, fought against two enemy aircraft over Gorizia using his Mauser pistol. On 12 July Bosio, Resio and Calderara took off after sunset and daringly bombed enemy camps at Gorizia.

While it was to disband in late July, the 8ª Squadriglia moved on 6 August to Trevignano being assigned to the XIV Corpo d'Armata of the 3ª Armata. Up to 31 July its planes flew 49 reconnaissance missions, dropped 25 bombs and 3,000 darts and took 71 photographs. On 16 August Ripamonti *"took very useful photographs of the Castelnuovo-San Michele-Boschini line and suffered for a long time the fire of enemy artillery that hit his airplane several times."* For this mission he got a *Medaglia di Bronzo.* The Squadriglia went to Clauriano and was disbanded on 27 August, when it had flown 95 flights, 52 of them war missions. The *Medaglia d'Argento* honoured Resio, the Bronze Bosio, Calderara, Coniglio, and de Bratti.

A new 8ª Squadriglia with Nieuport biplanes was in formation at Mirafiori, Turin, since 26 July and then it moved to Aviano. The 8ª Squadriglia Nieuport Biplani consisted of commander *cap.* Guido Tacchini and pilots *capitani* Oreste Salomone and Matteo Scarpis, *tenenti* Domenico Bolognesi and Francesco Baracca. It was equipped with Nieuport 10 serials 384, 387, 388, 595 and 596.

During August the Nieuports were fitted with machine guns and the unit flew training flights. On 21 August two airplanes were moved to Campoformido for the defence of Udine, and on the 25th Baracca and *cap.* Scarpis flew the first war mission, taking off to intercept an enemy airplane.

On 9 September the whole Squadriglia moved to Santa

Left: *Cap*. Oreste Salomone, here in front of a Nieuport 10, was the first Italian airman to be awarded the Gold Medal for Valor.

Caterina, near Udine, and the same day there was the first combat. *Cap.* Bolognesi with Ni 387 and *ten.* Baracca with Ni 596 intercepted an enemy airplane raiding Udine. Baracca fired from close range but his gun soon jammed. Bolognesi fired 100 rounds at the enemy, then also *cap.* Tacchini joined in, pursuing the enemy. The Nieuport of *ten.* Baracca was hit in three places.

In the second half of October the unit received some French built Nieuport 10, numbered 1035, 1037 and 1039. On 19 November a raid against Udine by half a dozen Austrian aircraft had to face all the operative fighters of the 8ª Squadriglia: 386, 385, 1037 and 1035, with pilots Tacchini, Bolognesi, Scarpis and Baracca who fired a total of 473 rounds dispersing the enemy formation. One of the Austrians, Albatros B.1 24.17 of Flik 14, crashed on return to its base, the pilot *Kpr.* Veselic was unhurt but the observer *Leut.* Riesenecker was injured.

On 1 December 1915, according to the new structure introduced by the Comando Supremo, the 8ª Squadriglia Nieuport, classified as a fighter squadron, became 1ª Squadriglia da Caccia (Ni) and together with the 2ª Squadriglia da Ricognizione e Combattimento formed the *Gruppo delle Squadriglie per la difesa di Udine* (group of squadrons for the defence of Udine).

Victory Claims of the 8ª Squadriglia Nieuport

Date	Place	Pilots	Airplane	Victim
19.11.1915	Udine	*Ten.* Baracca, cap. Bolognesi	Nieuport	Albatros 24.17 Flik 24

Key to Abbreviations in Victory/Loss tables:

DOW: Died of wounds
KIA: Killed in action
KIC: Killed in combat
WIA: Wounded in action
WIC: Wounded in combat
UNH: Unharmed.

Above: Nieuport 10 #1037 armed with a Colt gun over the top wing and another unidentified gun on the side of the cockpit. The second airman is *cap*. Tacchini, commander of the 8ª Squadriglia, the third one is *ten*. Baracca.

Nieuport 10 Ni.387
***Cap*. Bolognesi**
8ª Squadriglia
September 1915

9ª Squadriglia

Above: Farman 268 of the 9ª Squadriglia, a cage is attached to catch the cartridge cases of the machine gun.

The 9ª Squadriglia Farman of the III Gruppo at the outbreak of war went to Pordenone assigned to the Comando Supremo. It was equipped with Farman 1914 with De Dion motor 268, 272, 274, 310 and 316. *Cap.* Giorgio Chiaperotti was its commander, the pilots were *cap.* Ivo Varanini, *s.ten.* Roberto Bonuti, *m.llo* Armando Sportoletti, with observer *cap.* Lelio Gaviglio.

In June it moved to Campoformido and flew reconnaissance missions over the mountains Sabotino and Kuc. On 7 June *s.ten.* Bonuti on Farman 274 dropped bombs for the first time. Again, on 13 June Bonuti bombed the railroad station at Pravcina starting a fire. In this period the squadron received smoke cartridges and lamps for signaling. On 29 June Bonuti again flew an artillery-spotting mission while hail ripped the fabric of his airplane's wings.

In mid-July the 9ª Squadriglia was assigned to the II Gruppo for the Second Battle of the Isonzo (18 July to 3 August), operating for the II Army Corps. On 15 July it bombed a paper factory at Podgora. It now had observers *ten.* Giulio Cavalli and *s.ten.* Alberto Cordero di Montezemolo, joined in August by *s.ten.* Giovanni Pirelli. On 22 July the joystick of the Farman of *s.ten.* Sportoletti broke, but the pilot leaned out of the nacelle, grabbed the control cables and managed to get his plane back on the ground. Then he immediately climbed in another plane and took off again. On 23 July Varanini and Cavalli were targeted by enemy rifle fire over Castagnevizza, and returned with a still engine, while Cordero di Montezemolo and Bonuti attacked with firebombs a kite balloon moored at Podlesce. At the end of July, the unit had flown 41 missions with 66 war flight hours.

On 9 August an enemy airplane attacked the Farman of Sportoletti and Cavalli over the Bainsizza, then their engine seized but they managed to land behind the Italian lines. In September the 9ª Squadriglia received Farman 269 and 454, still with De Dion engines, and pilots *serg.* Aurelio Sciarretta and *sold.* Euclide Benini. On 24 September Cordero di Montezemolo flew a daring mission over the airfield of Aisovizza. On 26 September the 9ª Squadriglia, now with *cap.* Varanini as commander, was definitely assigned to the II Gruppo. Being the only unit of this group, as the 7ª and 8ª Nieuport had been disbanded, it was very active in September on the sector of Aisovizza and Mount Korata.

On 10 October *cap.* Ermanno Beltramo became commander. Pilot *sold.* Umberto Ferro and *ten. oss.* Domenico Albertoni and Guido Pacinotti arrived. On 7 October the unit moved to S. Caterina for the coming offensive, and received Farmans with Fiat motor 531, 539, 549, 553, 558, 564 and 573 with which a 3ª Sezione was formed. On 19 October it carried an exceptional observer, the poet Gabriele d'Annunzio who flew with Beltramo in a reconnaissance to Chiapovano. Again, d'Annunzio and Beltramo flew for two hours over the Bainsizza on the 22nd.

Above: Farman 553 of the 9ª Squadriglia at S. Caterina airfield, near Udine.

Above: Farman 539 of the 9ª Squadriglia showing the fins faired with cloth, and camouflage paint applied to the horizontal tail and to the top wing. (Archive Caliaro)

On 23 October the unit flew seven reconnaissance missions, and nine the next day, including one more by Beltramo and d'Annunzio. On 25 October the unit flew twelve war and test flights.

In November 1915 new Farmans arrived, including serials 799, 810 and 881. On 23 November *serg.* Filippo Palange and *s.ten.* Pirelli pursued an enemy plane flying over Udine, the same day *ten.* Albertoni braved enemy fire over Aidussina. On 1 December 1915 the unit was named 9ª Squadriglia da Ricognizione e Combattimento, then during the winter it scaled down its activities on account of bad weather.

On 14 December three airplanes with pilots Beltramo, Palange and Sportoletti attacked Austrian camps at Chiapovano dropping bombs and steel darts and then they went down to strafe the enemy troops. On 21 December a formation of four aircraft bombed enemy positions on the Bainsizza, with pilots Beltramo, Sportoletti, Palange and Benini. At the end of the year *s.ten.* Stefano Fedeli became temporary commander. During 1915 the *Medaglia d'Argento* was assigned to *tenenti* Albertoni and Cavalli and to *capitani* Beltramo, Chiaperotti and Varanini, *Medaglia di Bronzo* to Cavalli, Cordero di Montezemolo (two) and *m.llo* Sportoletti.

On 1 January 1916 at the command of *s.ten.* Fedeli the 9ª Squadriglia included pilots *s.ten.* Sportoletti, *sergenti* Benini, Palange and Sciarretta, observers *tenenti* Albertoni and Pacinotti, *s.tenenti* Cordero di Montezemolo and Pirelli. On 1 February *cap.* Luigi Chiappelli became the new commander. On 10 February *s.ten.* Stefano Fedeli with *ten.* Pirelli had a combat with an enemy airplane that was driven off. On 28 March 1916 the 9ª Squadriglia was passed to the Gruppo Squadriglie per Artiglieria with the new name 7ª Squadriglia Aeroplani per Artiglieria and moved to Oleis.

Left: Farman 810 of the 9ª Squadriglia.

10^a Squadriglia

Above: Farman 273 of the 10^a Squadriglia. On the fuselage there is a plaque with the name of this airplane, "*F. Crispi*", an Italian prime minister of the XIX Century.

Formed on 1 April 1913, the 10^a Squadriglia was part of the III Gruppo and it was mobilized at Pordenone, coming from Taliedo, Milan, on 23 May 1915. It was equipped with Maurice Farman 1914 with De Dion 80 hp engines, among them 273, 306 and 312. Its aircraft, paid for by a public subscription, carried individual names such as "Irpinia", "Novara", "Ferroviere" and "Francesco Crispi". *Cap.* Carlo Tappi was the commander, with pilots *capitani* Luigi Bailo and Amedeo Ferraro, *marescialli* Domenico Perucca and Enrico Burzio, and observer *cap.* Oliviero Rodolfo.

On 25 May there was the first war mission: two airplanes attacked the station of Gorizia. The unit moved to Campoformido at the end of May, and got *ten. oss.* Armando Giammarco. From 28 to 31 May it was detached at Bazzera because an enemy air raid against Venice was expected, then it went to Campalto, also close to Venice, with one flight at Chiasiellis and one at Borgnano. On 3 June it returned to Campoformido. On 6 July the Squadriglia bombed Aisovizza airfield and the station of Nabresina. On 11 July it moved to S. Maria la Longa, assigned to the III Gruppo. It received for a period some pilots of the 11^a Squadriglia that was changing its aircraft. During the 2nd Battle of the Isonzo, 18 July to 3 August, it was based at Medeuzza.

On 13 September the III Gruppo was disbanded and the 10^a Squadriglia was attached to the I Gruppo, remaining at S. Maria la Longa. With *cap.* Tappi there were at this time *ten.* Cesare Sacerdoti, *s.ten.* Mario Ponis, *m.llo* Perucca, *serg.* Gualberto Giuntoli, and *tenenti osservatori* Mario Ajmone Cat and Giuseppe Ottolenghi.

On 11 October Farman "*Irpinia*" was lost, shot down over Oppachiasella during a reconnaissance and bombing mission to Castagnevizza, but pilot *serg.* Giuntoli and *ten.* Ottolenghi survived and brought back the photo plates from the wreck of the airplane on the banks of the Isonzo. On 16 October the command of the VII Corpo d'Armata praised this unit.

On 7 November 1915 the airplane of *cpr.* Roberto De Anna and *ten.* Ottolenghi was destroyed by the explosion of one of its own bombs while landing, but the crew was unharmed. On 1 December the unit became 10^a Squadriglia da Ricognizione e Combattimento. In 1915 it had flown 211 war missions.

On 1 January 1916 the 10^a Squadriglia had in force *cap.* Tappi, *s.tenenti* Enrico Burzio and Mario Ponis, *asp.* Perucca, *sergenti* De Anna and Giuntoli, *cpr.* Carlo Adamoli, observers *capitani* Guido Barutta and Oliviero Rodolfo, *tenenti* Carlo Lombardi, Giulio Roberto Pastore and Ferruccio Capuzzo. In early 1916 the unit was very active above the lower Isonzo and the coast. On 9 January there was the first action of the

Above: Farman 306 of the 10ª Squadriglia, the plaque is named *"Bari"*, a city in Southern Italy.

year, a reconnaissance along the coastal road from Sistiana to Opicina. On 15 January there was an air combat: Perucca and Rodolfo, coming back from a reconnaissance, pursued an enemy aircraft toward Aidussina, then they had a second duel with a plane west of Gorizia. On 29 January *serg.* De Anna and *ten.* Capuzzo attacked and pushed back an enemy airplane that tried to cross the lines.

On 19 February a Fokker above Oslavia attacked Rodolfo and Burzio who had to use a pistol for defense. The Fokker harassed them a long time, then left them alone. On 19 March Capuzzo bombed the Borgo Carinzia station of Gorizia. The same day another plane with *serg.* Adamoli had a duel with a Fokker near Gorizia and fired 60 rounds with its machine gun. On 31 March *cap.* Tappi with *magg.* Gallina, commander of the Gruppo, flew to Gorizia and dropped a message reporting the funeral rites for Major Wladimir Junovicz, a victim of the luckless Austrian raid of 27 March that caused the loss of five aircraft and the capture of five airmen. On 4 April an Austrian plane attacked *serg.* Forestieri above Cervignano, the Italian fought back firing no less than 290 rounds and saw the Austrian gliding away toward Gorizia.

On 7 April the 10ª Squadriglia was shifted to Villaverla, near Vicenza, on the northern front, assigned to the 1ª Armata. On 15 April it became the 27ª Squadriglia Farman. In this period the Silver Medal went to *asp.* Perucca and to *ten. pil.* Sacerdoti, the Bronze to *m.llo* Burzio, *cpr.* De Anna and *cap. oss.* Barutta.

Above: A Farman of the 10ª Squadriglia with a pipe projecting from the nacelle, possibly a device to drop bombs.

Above: *Ten*. Ferruccio Capuzzo, an artillery officer, became a valiant commander in the war; he was killed in action in Libya in 1925. A fort was named after him, where the first battles in North Africa were fought in 1940.

11ª Squadriglia

Above: Farman Type 1912 serial 207 of the 11ª Squadriglia before the war. (Archive Caliaro)

Created in 1914 and equipped with Farman 1912, at the outbreak of war the 11ª Squadriglia remained at Brescia, outside of the war zone, to convert on the Maurice Farman 1914. At that time its commander was *ten.* Michele Masi, but in June commander was *cap.* Mario Girotto with pilots *ten.* Masi, *s.ten.* Luigi Orlando, *serg.m.* Prospero Freri (the future designer of Italian parachutes). On 3 July the unit got its new aircraft: Farmans 458, 461 and 462, so on 7 July it moved from Brescia to La Comina, entering the I Gruppo to replace the disbanded 3ª Blériot, getting also the observers of the latter: *tenenti* Armando Fabretti and Enrico Giorgetti.

In the night of 15/16 July, according to orders, it changed the national markings, removing the black circles and bars on the wings, and applying red, white and green on rudders and wings undersurface.

The Squadriglia moved to Chiasiellis and began operations on 18 July flying a reconnaissance to Redipuglia. The next day it bombed with incendiary devices a kite balloon at Castagnevizza. Its Farmans also carried *cap.* Novellis, commander of the I Gruppo, and *ten.* Tullio Toccolini, who was transferred from the 13ª on 26 July. On 31 July from the disbanded 5ª Nieuport came *ten.* Giuseppe Rossi and Ercole Messi, and then *ten. oss.* Gastone Novelli. On 1 August an enemy aircraft attacked the plane of Orlando and Giorgetti over Monfalcone. On 6 August Farman 461 ran out of fuel and crashed, so the 11ª remained with just two planes. On 10 August *cap.* Masi was appointed commander of the 2ª Squadriglia, from which came to the 11ª *soldato pilota* Vittorio De Lucchi, and to which went *ten.* Messi and Giorgetti (who returned on the 22nd). Finally on 17 August the observer *ten.* Michele Teodosio De Bonis and pilot *cpr.* Agostino Consigli also arrived.

Waiting for re-equipment, the 11ª passed its pilots (in August it was the only squadron of the I Gruppo) to support the 9ª and 10ª. In September this Squadriglia was at Chiasiellis in the I Gruppo of the 3ª Armata, but operationally reporting to the II Gruppo of the 2ª Armata. On 10 September *s.ten.* Orlando became temporary commander: four days later the effective commander arrived, *cap.* Francesco Vece, but on that very day he crashed on landing Farman 318 and was injured, while his passenger *ten.* De Bonis died. For the rest of September, no flying was done, as the unit was out of aircraft.

On 1 October *cap. pil.* Ivo Varanini took over command, and *s.ten. oss.* Cesare Franceschini, *sold.* Francesco Dossi and *cap.* Toccolini joined the unit. On 8 October Squadriglie 7ª, 10ª and 11ª flew out to attack the house used as the command post of Archduke Joseph August at Castagnevizza, but the two Farmans of the 11ª had to abort the action. On 16 October the unit deserved a commendation from the command of the VII Corpo d'Armata.

The 11ª Squadriglia took part in the 3rd Battle of the Isonzo, since 18 October, attacking the enemy with bombs and darts. On 30 October the airplane of Franceschini and *sold. pil.* Dossi was wrecked at Borgo San Valentino, with no harm for the two airmen. The 11ª Squadriglia was left with just Farman De Dion 273, then it received Farman Fiat 562 and 806. On 16 November it took on strength *serg. pil.*

Right: Farman MF 461 of the 11ª Squadriglia. (Archive Caliaro)

Above: Carlo Adamoli designed the unusual Adamoli Cattani fighter. In 1934–35 he flew across Africa looking for rare metals, and died in 1942 flying a Ju 87 Stuka.

Antonio Maramaldo della Minerva and Farmans 817 and 857. On 25 November during the 4th Battle of the Isonzo an Austrian aircraft, identified as an Aviatik, (actually Fokker 03.51 of *Oberlt*. Bernath of Flik 4) shot down the Farman of Franceschini near S. Lorenzo di Mossa, wounding the pilot. The airplane on the ground was wrecked by artillery fire but Italian soldiers daringly recovered what was left. On the 28th the airplane of Dossi and Toccolini crashed due to engine failure at Borgnano, the airmen survived. In a post war book, *ten.* Rossi described a spotting mission for the 305-mm guns of the Navy when his Farman, with pilot *serg.m.* Freri, was shot down by Austrian guns over Doberdò and crashed behind Italian lines, but the date of this event is unknown.

On 3 December 1915 the unit, now 11ª Squadriglia da Ricognizione e Combattimento, received pilots *caporali* Alfredo Ghidotti and Giuseppe Alfonsi, and the observer *ten.* Salvatore Travaglianti. During 1915 this unit had flown 6 standing patrols, 3 bombing and 44 reconnaissance missions.

In January 1916 the 11ª Squadriglia had two aircraft and pilots *cap.* Varanini, *s.ten.* Orlando, *serg.* Freri, *caporali* Dossi, Ghidotti, Alfonsi, observers s.*tenenti* Franceschini and Novelli, *ten.* Travaglianti and *cap.* Toccolini. It flew in a large operation on 18 February with all the units of its sector, then it didn't fly from 27 February to 13 March. On 20 March Toccolini left, and *s.ten. oss.* Giovanni Pirelli arrived, and five days later *ten.* Ugo Niutta. On 31 March *cpr.* Giuseppe Patrito with gunner *soldato* Ercole Patrucco had a combat over Udine against a fast enemy airplane, but their gun jammed. On 4 April Travaglianti and Maramaldo had a 15 minutes combat over Mount Sei Busi, their Farman was riddled with bullets but they escaped. Their likely enemy was

Above: Farman MF 845 of the 11ª Squadriglia.

Brandenburg 61.59 of Flik 2 piloted by *Leut*. Bogun Buriat. On 6 April the unit dropped bundles of newspapers.

On 8 April *cap.* Enrico Amour replaced Varanini as commander. The 11ª became 28ª Squadriglia on 15 April 1916. *Medaglie d'Argento* were assigned to Maramaldo della Minerva and Cesare Franceschini, Bronze to Patrito.

12ª Squadriglia

Above: A Farman of the 12ª Squadiglia in the first days of the war, still with the black circle and bar markings. (Archive Caliaro)

Equipped with Farman model 1912, the 12ª Squadriglia at the declaration of war remained at Verona to convert onto the Farman 1914 80 hp. *Cap.* Ferdinando De Masellis was its commander, its pilots were *ten.* Ferdinando Bonazzi, *s.ten.* Enrico Buzzi, *m.llo* Silvio Picolla, and it had Farman 46, 262, 264 and 265 with Renault 70 hp motor, and 63, 313 and 914 with De Dion motor, which it received on the first day of war, May 24, 1915.

On 6 June, mobilized, it went to the northern front at Asiago operating for the 1ª Armata. It was not attached to any Gruppo, and at the front its pilots were De Masellis, Bonazzi, *ten.* Mario Girotto, *m.llo* Ettore Bosco and *serg.m.* Nicola Valentino, observers *s.ten.* Giuseppe Maceratini and Pietro Pini.

On 12 June *cap.* De Masellis and *ten.* Pini tried to fly the first war mission but their plane couldn't climb, so it was *ten.* Bonazzi who flew the first reconnaissance with Farman 316. Two new Farman De Dion, 459 and 460 arrived on that day. The next day Bonazzi took off with *s.ten.* Forza, but he fell down from a height of 30 feet, proving that the aircraft of this unit lacked the power to operate on that high plateau region. On June 26 five reconnaissance missions were flown. On 30 June the 12ª Squadriglia sent out *ten.* Bonazzi and *sold.* Battaglini on Farman 316 to intercept the first Austrian bombing operation against Asiago.

On 2 July 1915 Farman 459 crashed and Bonazzi was slightly injured. On 15 July Bonazzi took over command from De Masellis who had another crash with Farman 313. In July the unit took the first photographs of enemy territory and tested communications using a Donath lamp for optical telegraphy.

In August the squadron flew De Dion-engined Farmans 316, 452, 460, 492 and 499. It was assigned to the Command of Artillery of the V Corpo d'Armata. It flew some patrol flights to intercept enemy aircraft, a pointless exercise considering how totally underpowered the Farmans were. In mid-September the 12ª began receiving Farmans with Fiat 100 hp engines, the first one was 534 flown in by *cap.* Ermanno Beltramo.

On 20 September 1915 *cap.* Beltramo flew over Trento, the provincial capital that was, with Trieste, one of the goals of Italy's war, carrying the poet Gabriele d'Annunzio who dropped patriotic messages on the city. In October the unit had Farmans with Fiat engine 534, 802 and 803.

In November the unit was asked to provide difficult interception missions against Austrian raids and in one of these, on 7 November enemy fire damaged the Farman 534, of *cap.* Bonazzi. That day two other Farmans crewed by *cpr.* Giulio Adelasio with *ten. oss.* Maceratini, and *serg.* Cruciani with *ten. oss.* Antonio Stefani, attacked two enemy planes that flew away. *Serg.* Alessandro Borgeto, on 14 November, crashed Farman 548 due to an engine failure but he survived as did his observer *ten.* Stefani. On 20 November 1915 enemy rifle fire hit the airplane of Adelasio and Maceratini flying a reconnaissance over Folgaria, the observer was wounded in a leg but he didn't stop his mission. The courage and efficiency of this squadron were remarkable, as it operated in a harsh mountain environment.

Above: Farman MF 452, 316, and 451 of the 12ª Squadiglia at Asiago. (Archive Caliaro)

Above: Gabriele d'Annunzio in the front seat of Farman 539; two bombs are attached under the bottom wing.

Gabriele d'Annunzio and *cap*. Beltramo after their flight over Trento on 20 September 1915. (Archive Caliaro)

Farman 802 of the 12ª Squadiglia in a photo signed by gunner Battaglini (possibly Cesare Battaglini, later a SVA pilot). (Archive Caliaro)

Above: A bad landing for Farman 817, probably of the 12ª Squadiglia, at Villaverla.

On 1 December 1915 the unit became 12ª Squadriglia da Ricognizione e Combattimento and went to the reborn III Gruppo, moving on the 12th from Asiago to Villaverla. On 13 December Bonazzi went to command the 1ª Squadriglia and passed command of the 12ª to *cap.* Armando Armani. From 24 May to the end of 1915 the 12ª flew 137 war flights. *Ten.* Antonio Gallotti and *ten.* Maceratini earned the Silver Medal, *cpr.* Adelasio, *cap.* Bonazzi and *ten. oss.* Pini the Bronze Medal.

By 1 January 1916 the 12ª Squadriglia, based at Villaverla, had on strength the pilots *cap.* Armani, *ten.* Antonio Gallotti, *serg.* Cruciani, *cpr.* Adelasio, *cpr.* Borgato, *cpr.* Francesco Carabelli, and the observers *ten.* Maceratini, Poli and Stefani, being equipped with Farmans 534, 802, 803, 804, and 814.

The year's actions began on 10 January when Gallotti and Armani bombed the station of Trento. On 16 January the Farman of Cruciani and Armani attacked an Austrian command in the Lagarina Valley, and was repeatedly hit by ground fire. On 28 January *cpr.* Carabelli with a gunner had a combat with an enemy aircraft.

On 4 February a 3ª Sezione of the Squadriglia was formed at Verona, equipped with Caudrons, with pilots *s.ten.* Pietro Cavadini, *soldato* Andrea Teobaldi and *soldato* Giovanni Conti, with the observer Carlo Emanuele Parini who was later assigned to the 1ª Squadriglia.

The Caudrons were equipped with a wireless set for communicating with the artillery, while the Farmans only used smoke devices or lamps. On 10 February three airplanes bombed camps at Calliano, on 15 and 20 March barracks near Luserna were hit, on 4 April the squadron bombed barracks at Levico. On 9 April three Austrian planes attacked the Farman of Carabelli and *ten.* oss. Carlo Lombardi, even using steel darts (*Zgsf.* Hablitschek and *Leut.* Terk of Flik 7 claimed an enemy plane forced down).

The 12ª Squadriglia became 32ª Squadriglia Aeroplani on 15 April 1916. *Serg.* Cruciani received a *Medaglia d'Argento.*

13ª Squadriglia

Above: Blériot 246 of the 13ª Squadriglia.

The 13ª Squadriglia Blériot at the outbreak of war belonged to the I Gruppo; it moved from Aviano, where it was based since January 1915 at Torresella, on 21 May 1915, where it was assigned to the III Gruppo. Its commander was *cap.* Vincenzo Lombard, with pilots *ten.* Luigi Carnevale, *s.ten.* Bartolomeo Costantini, *sergenti* Dante Binda, Giuseppe Rolando and Aurelio Sciarretta, flying Blériots 39, 246, 250, 284, and 297.

On 26 May 1915 Lombard dropped the first bombs on Gorizia. On 29 May the 13ª moved from Torresella to Pozzuolo del Friuli. Many reconnaissance missions were flown with observers *tenenti* Armando Fabretti and Tullio Toccolini. On 16 July the unit received *soldato* Silvio Amico with Blériot 300. The order to disband this squadron was issued on 25 June 1915, but the 13ª remained operative until 6 July when the Blériot of Lombard came back to base badly hit after a bombing raid. The 13ª Squadriglia was disbanded on 1 August 1915, it had flown 38 war missions, including two bomb raids.

On 2 March 1916 a new 13ª Squadriglia da Ricognizione e Combattimento per l'Albania was reformed at Taranto. Its commander was *cap. pil.* Leopoldo De Rada, pilots *ten.* Manfredo Calosci, *serg.* Oreste Brambati, *cpr.* Oreste Ciapetti, *sold.* Lorenzo Barberis, observers *cap.* Fausto Pesci, *ten.* Carlo Ammazzalorsa and *s.ten.* Augusto Gauthier, with six Farman Fiat 100 hp.

On 3 March the unit sailed to Valona, Albania, and set base at Saline. Assigned to the XVI Corpo d'Armata in Albania, it soon moved to a more manageable field and began war flights on 17 March.

On 1 April personnel of this unit took part in the landing at Punta Samana: two Macchi Lohner of the Italian Navy with pilots *Tenente di Vascello* Roberti and *S.ten. di Vascello* Lionello Caffaratti, with *cap.* De Rada, *cap.* Pesci and two motorists alighted near an Austrian base and destroyed some installations and deposits of coal. On 14 April *cpr.* Ciappetti and *cap.* Pesci flew a long mission of 3 hours and 20 minutes, then they force landed for engine troubles, during the night strong winds destroyed their Farman.

The 13ª became 34ª Squadriglia on 15 April 1916.

Left: Blériot 297 with two airmen on board ready for take-off.

Right: The airmen after the successful raid at Punta Semana, TV Roberti, *cap*. De Rada, STV Caffaratti, and *cap*. Pesci.

14ª Squadriglia

Right: Blériot 43 of the 14ª Squadriglia. (Archive Achille Vigna)

The 14ª Squadriglia Blériot, in May 1915 was part of the I Gruppo and on 21 May it moved from Aviano to Portogruaro Torresella with Blériots 7, 42 ,43, 44 and 252. It was made available to the III Gruppo until 28 May, when it moved to Chiasiellis. The airmen of the 14ª Squadriglia were *cap.* Tullio Visconti, pilots *cap.* Carlo Graziani, *ten.* Ercole Ercole, *serg.m.* Aurelio Sciarretta and observer *ten.* Tullio Toccolini.

This unit flew mainly reconnaissance missions. On 13 June Ercole, Graziani and Visconti bombed a livestock park at Doberdò. This squadron was disbanded on 22 June having flown 25 war missions with four bomb raids. Graziani had earned one *Medaglia d'Argento* and a *Bronzo*.

A new 14ª Squadriglia Farman was formed, that in August 1915 got the name of 2ª Squadriglia Farman.

15ª Squadriglia

Left: A Caproni Parasol at the factory, still without its fabric covering.

In 1914 the 15ª Squadriglia was equipped with Blériots, based at Piacenza in support of the local siege artillery park. It is not reported as existing in January 1915, but a new 15ª was created, equipped with Farman 1912, with TV Ugo De Rossi as commander. 15ª Squadriglia in April 1915 converted to the Caproni 80-hp monoplane, having airplanes 228, 229, 230, 231, 232, 233, 234. Assigned to the III Gruppo, at the outbreak of war it remained at Piacenza. Its personnel included commander *cap.* Umberto Rossi, pilots *ten.* Gaetano Turilli, *s.ten.* Ettore Croce, *brig.* Domenico Cattaneo, *serg.* Virgilio Sala. On 24 June it moved to Pordenone for the Parco d'Assedio di Artiglieria (artillery siege park) and to convert on the Caproni 2 Parasol 100 hp. The squadron flew many reconnaissance training flights with observers *tenenti* Casoria, Gozzano and Maceratini, starting on 8 July, and tested the Parasol and aerial bombs. As the Parasol was prone to accidents, one of which caused the death on 20 August of *brigadiere* Domenico Cattaneo when a wing broke in flight, the 15ª Squadriglia was disbanded on 15 September 1915.

Above: The Caproni Parasol now exhibited at Omaka, New Zealand. (Wikimedia Oren Rozen)

16ª Squadriglia

Very little is known about this unit. The official Chronology of the Italian Air Force does not report its existence in December 1914, nor among the mobilized units on 24 May 1915 but it was reported existing in January 1915 at Piacenza, with temporary commander *ten.* Matteucci, *tenenti* Ercole, Stanzani, Oreste Salomone and *serg.m.* Arturo De Bernardi. It was based at Piacenza, assigned to the local Reggimento d'Assedio (Regiment of siege artillery). It is known to have received Blériots serials 1, 7, 24, 38, 40, 247, 248, 280 but it was immediately disbanded.

17ª Squadriglia

This unit was formed on 23 November 1913 at San Francesco al Campo, near Turin, and shipped to Tripoli, Libya, equipped with Farmans under the command of *ten.* Giorgio Chiaperotti with pilots ten. Andrea Poggi and Carlo Tappi, *m.llo* Domenico Perucca, *serg.m.* Enrico Burzio. In 1914 it planned a postal service from Tripoli to Tahruna and offered transport flight for officers, that found only one taker, a Major doctor. The unit also organized a successful course for observers. On 2 June 1914 17ª Squadriglia was disbanded and its men and equipment returned to Italy.

Bleriot XI BL.251
1[a] Squadriglia
Spring/Summer 1915

Bleriot XI BL.286
2[a] Squadriglia
Spring 1915

Bleriot XI BL.283
3[a] Squadriglia
1915

Bleriot XI BL.246
13ª Squadriglia
1915

Nieuport IV Ni.111
6ª Squadriglia
1915

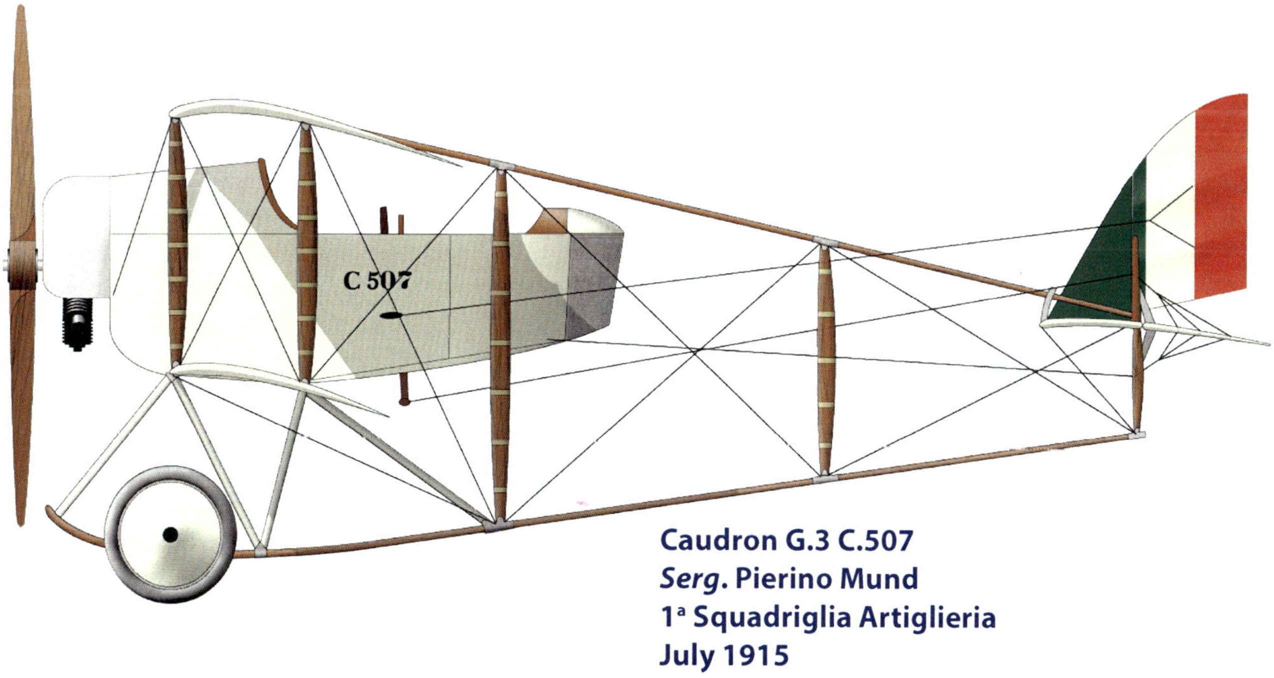

Caudron G.3 C.507
***Serg.* Pierino Mund**
1ª Squadriglia Artiglieria
July 1915

Squadriglie per l'artiglieria (Artillery Squadrons)

1ª Squadriglia per l'Artiglieria

Above: The Caudron G.3s of the 1ª Squadriglia Artiglieria being prepared for action at the airfield of La Comina, Pordenone.

The 1ª Squadriglia Artiglieria was formed on 26 June 1915 with Caudron G.3s at Pordenone. On 2 July it went to Medeuzza in the I Gruppo Squadriglie di Artiglieria, assigned to the 2ª Armata. It was equipped with Caudrons serials 1, 346, 363, 373, 500 and 501. It was ordered to operate for the II and VI Corpo d'Armata. Its personnel included the commander *cap. oss.* Emilio Gamerra, pilots *cap.* Filippo Sisto, *s.ten.* Arturo Clerici Bagozzi, *sergenti* Vincenzo Arone, Dante Binda, Mario Guida, Mario Pez and Giuseppe Rolando, the observers *cap.* Felice Porro, *tenenti* Vincenzo Fumero, Aldo Silvestri and Vincenzo Velardi. Porro was the commander of the Regia Aeronautica in North Africa in 1940-41, in 1935 he wrote a book, "*La Guerra nell'aria*", an excellent story of the Italian Air Force in the Great War.

On 11 July, during a ferry flight Caudron G.3 507 with pilot *serg.* Pierino Mundo got lost and ended up in Austrian land. The pilot was captured and the airplane was later used in operations with Flik 4, now serialled 00.06. The 1ª Squadriglia d'Artiglieria began operations in July during the offensive against Mount San Michele. On 4 August it moved from Medeuzza to Bolzano del Friuli, (not to be confused with Bolzano, Bozen, South Tyrol) and on 7 September from Bolzano to Oleis. Meanwhile it received *ten.* Alvaro Leonardi and G.3 364. On 22 October 1915, a Caudron crashed on Oleis airfield, seriously injuring *s.ten.* Giuseppe Zanchi and *ten. oss.* Napoleone Del Duca, who had tried to fly a mission despite a strong wind. During 1915 this squadron flew 228 war missions. The *Medaglia d'Argento* rewarded *ten.* Del Duca and the *Bronzo cap.* Porro.

On 1 January 1916 the 1ª Squadriglia Artiglieria with commander *magg.* Gamerra had pilots *cap.* Arturo Oddo, s.*tenenti* Clerici Bagozzi, Umberto Guazzoni and Giuseppe Retinò, *sergenti* Arone, Binda, Rolando, Pez, *cpr.* Giuseppe Garuffo and *sold.* Luigi Vulcano, observers *cap.* Marc'Aurelio Barbarisi, *tenenti* Giglio and Velardi, s.*tenenti* Michelangelo Caetani, Fabio Fabi, Ranieri Sabatucci and Silvestri.

On 15 January *s.ten.* Sabatucci and *sold.* Vulcano were seriously wounded in the crash of their airplane hit by enemy fire. On 3 March Gamerra passed his command to *cap.* Raul Lampugnani, former commander of the 5ª Caudron.

In mid-March the unit moved to Sammardenchia and then at the end of the month to Risano. On 15 April 1916 it got the new designation 41ª Squadriglia. At that time one more *Medaglia d'Argento* had gone to Gamerra and a Bronze to Guazzoni.

Casualties and Combat Losses of the 1ª Squadriglia Artiglieria

Date	Place	Aircraft	Crew	Cause
9 or 11.7.1915	Gorizia	Caudron G.3 507	*Serg.* Mundo POW	Accident
24.10.1915	Oleis	Caudron	*S.ten.* Zanchi WIC. Ten. Del Duca WIC	Accident
15.1.1916	Bodigoi	Caudron	*Sold.* Vulcano WIA, ten. Sabatucci WIA?	Ground fire?

Above: Austrian servicemen posing with the brand new Caudron G.3 507 which a confused pilot had brought to them.

Right: Women working on Nieuport wings at the Macchi factory in Varese.

2ª Squadriglia per l'Artiglieria

Above: A Macchi Parasol at the factory; in the rear seat is Macchi's test pilot Clemente Maggiora.

The 2ª Squadriglia Aviatori per Artiglieria equipped with Macchi Parasols was formed on 26 June 1915 at Pordenone, and it moved to Medeuzza in the I Gruppo Squadriglie di Artiglieria on 2 July, assigned to the 3ª Armata, with commander *cap. oss.* Mario Santi. It had Parasols with serials 393, 397, 399, 405, 406 and 408 and six pilots, two of them officers. Many airmen served in this unit in different times, often rotating among the other Artillery squadrons: pilots *sottotenenti* Giuseppe Grassa, Luigi Olivi, Natale Palli, *serg.m.* Giuseppe Screpanti, *serg.* Alessandro Piola Caselli, Carlo Frigeri, Roberto Ghelfi, Antonio Palpacelli, Bartolomeo Pognante and Antonio Cremonese, *cpr.* Ferruccio Guicciardi, Riccardo Ciotti, Goffredo Gorini, Dino Menegoni and Mario Stoppani (the future test pilot of Ansaldo and Cant). The many officers who flew as observers were *cap. oss.* D. Rolandi, *ten.* Barbalarga and Robotti, *s.ten.* Antonio Babini, Mario Beltrami, Arturo Bruni, Giovanni Gallo, Armando Del Sole, Fabio Jannoni Sebastianini, Antonio Ottelio, Giancarlo Massola and Attilio Zoppi.

The bad qualities of their airplane are revealed in the citations of the medals assigned to the airmen of this unit. To *s.ten. oss.* Bruni: *"After four serious accidents, from which he luckily escaped unharmed, he went on flying against the enemy, despite the absolute rest ordered by the doctor on account of his nervous shock."* The citation for *cap.* Santi himself: *"Commander of an aviation squadron for the artillery, he was a good example and spur to the pilots and observers of the squadron, daringly facing serious flying problems and enemy offences, gaining important results from reconnaissance service"* And *ten. oss.* Gallo *"... kept on flying in bad weather conditions with a type of airplane of difficult control, with which he already had a dangerous accident."*

In November the 2ª Squadriglia converted to the much safer Caudron, replacing the Macchi that justified a medal to those who just dared to fly it. Both planes used the same Le Rhone 80 hp engine. At the end of the year the unit moved to Chiasottis-Risano airfield (villages between Udine and Palmanova). During 1915 247 war missions were flown, with active participation in the attack at S. Michele in July and then in the October Battle of the Isonzo. In that month the 2ª Squadriglia Artiglieria gave to the 3ª seven pilots and three observers. In 1915 a *Medaglia di Bronzo* was tributed to *s.ten.* Bruni, *s.ten.* Gallo and *serg.* Pognante.

In early 1916 Santi passed command to *cap. oss.* Marc'Aurelio Barbarisi. From 19 to 21 January *cpr.* Cremonese and *ten.* Gallo flew daring missions over the Carso returning with their plane hit by enemy fire. In January the unit flew 27 visual and photo reconnaissance missions and five artillery spotting, in February seven reconnaissance and ten spotting flights. *Ten.* Gallo again on 8 and 9 April

Macchi Parasol 348 carrying the pre-war eagle shield on its rudder.

Macchi Parasol 406: apparently, it could fly without its engine cowling.

Above: Personnel of the 2ª Squadriglia Artiglieria in a well-arranged souvenir photo.

flew daring photo reconnaissance missions despite enemy artillery, and another outstanding observer was *ten. oss.* Manlio Molfese (who after the war wrote the history of Italian air reconnaissance).

The 2ª Artiglieria flew 9 reconnaissance missions in March and 44 in April, plus six gun spottings. In this period Gallo got the *Medaglia d'Argento* and *Bronzo*, while *cap.* Santi, *serg.* Ciotti, *serg.* Guicciardi, *cpr.* Cremonese and *ten. oss* Molfese each had a Bronze. On 15 April 1916 the 2ª Artiglieria became 42ª Squadriglia.

Above: Macchi Parasol of the 2ª Squadriglia Artiglieria at Medeuzza. The rudder now carries the national colors, but the Savoy eagle has been preserved. (Archive Longhi)

A Macchi Parasol at Medeuzza, showing the panel added on the top of the wing to improve stability.

Above: A Macchi Parasol with a different front fuselage, with modified struts and engine cowling.

3ª Squadriglia per l'Artiglieria

Above: A Macchi Parasol fitted with the stabilization panel over the wing. (Archive Caliaro)

Formed at Taliedo on 22 September 1915, the 3ª Squadriglia Aeroplani per l'Artiglieria equipped with Macchi Parasol 401, 408, 413, 415, 416, 418, 419, 421, 422, was mobilized on 24 September at Medeuzza, in the I Gruppo Squadriglie Artiglieria, assigned to the Artillery Command of the X Corpo d'Armata of the 3ª Armata, covering the area between San Martino and Doberdò. Its commander was *cap.* Giulio Maurel, pilots were *s.ten.* Giuseppe Grassa, *sergenti* Roberto Ghelfi, Bartolomeo Pognante, Giuseppe Screpanti, *soldati* Goffredo Gorini and Dino Menegoni, and observers *ten.* Armando Del Sole, Mario Beltrami and Fabio Jannoni Sebastianini, *sottotenenti* Arturo Clerici Bagozzi and Amedeo Raffone. Of these, Grassa, Pognante, Ghelfi, Menegoni and Gorini, and the observers Del Sole, Beltrami and Jannoni came from the 2ª Artiglieria while Clerici and Raffone came from the 1ª Artiglieria.

It then received *cap.* Lampugnani and *serg.* Frigeri, and from the 1ª Sq. Art. *cap. oss.* Porro. In operation the 3ª was quite effective in spotting for the guns, having developed good forms of co-operation, but its Macchi Parasols were overly troublesome, so the squadron converted on the Caudron G.3, receiving nine of them on 29 November, known serials are 1057 and 1058. Also in this case, the dangerous features of the Parasol are reflected by the citations of the medals: to Stoppani: *"Despite having had many flying accidents, he kept on piloting a model of airplane of difficult control, disregarding bad weather conditions and the strong enemy ground fire, that once hit his airplane in seven parts and him too, though not wounding him."*

The 3ª Squadriglia was tasked to operate also for the XIII Corpo d'Armata. At the end of the year enemy cannons targeted the airfield of Medeuzza and the Squadriglia moved back to Gonars, a few miles west, then it returned to Medeuzza.

During 1915 the unit flew 56 war missions, all low-level reconnaissance flights, during which enemy fire often hit the aircraft, some of which resulted in forced landings. *Medaglia d'Argento* was assigned to *caporale* Mario Stoppani, and Medaglia di Bronzo to *ten.* Beltrami, *ten.* Del Sole, *ten.* Jannoni, *cap.* Porro, *serg.* Pognante and *serg.* Screpanti.

On 9 January the Caudron of *cpr.* Mario Sereni and *ten. oss.* Pietro Baggio was lost in action, with no injury for the two airmen. In February 1916 the 3ª Squadriglia moved to Gonars. In early April Maurel left to become commander of 8ª Artiglieria and *cap.* Porro became commander. A daring reconnaissance over the Carso on 1 April 1916 earned *ten.* Ferruccio Ranza a *Medaglia di Bronzo*, but his Caudron G.3 was badly hit and written off. The 3ª Artiglieria on 15 April 1916 became 43ª Squadriglia.

Above: A Macchi Parasol being delivered on a truck shows the early insignia of black circle and bar.

Above: A Macchi Parasol flying with the bottom part of the engine cowling removed.

Medeuzza airfield, with a Caudron G.3 in the background; the big arrow was used to show the direction of the wind to the pilots.

4ª Squadriglia per l'Artiglieria

Above: *Sottotenente osservatore* Pietro Ottelli with a Caudron G.3. (Archive Schifano)

The 4ª Squadriglia Artiglieria was formed in September 1915 at the Centro di Formazione of Ghedi with Caudron G.3s, and it was activated on 5 October 1915. Deployed at Gonars for the artillery of the VII Army Corps of the 3ª Armata, the 4ª Squadriglia had commander *cap. oss.* Giulio Costanzi with pilots *ten.* Fulco Ruffo di Calabria, *s.ten.* Franco Ambrosio, *sergenti* Pietro Botti, Silvio Scaroni, Raffaele Castellano, *soldati* Raffaele Ghislanzoni, Silvio Vernizzi and Filippo Volontè, later also *ten.* Ferruccio Ranza, and the observers *tenenti* Francesco Broili, Mario Lavalle, *s.ten.* Costantino Cattoi, Vincenzo Russo, Clemente Merli and later also *s.ten.* Achille Dander and Pietro Ottelli.

The unit began operations on 9 October flying a reconnaissance to Nabresina, and in October it flew 28 reconnaissance and 12 gun spotting missions for a total of 62 hours. On 21 October one of its airplanes flying at 1,000 meters was hit by anti-aircraft fire and forced to land near the Isonzo. On 7 November also the Caudron of *serg.* Scaroni with the observer *s.ten.* Cattoi was shot down by Austrian guns over Doberdò. On 9 November two planes flying at low level during a storm spotted Austrian guns shelling Italian infantry and using a Donath signaling lamp, directed effective counter-battery fire. On 15 November again the enemy flak hit the aircraft of *s.ten.* Ambrosio and *s.ten.* Cattoi. They had been flying low on account of fog and crash landed after being hit. In November and December bad weather reduced flights, just 18 reconnaissance and 19 gun spotting missions. During 1915 the *Medaglia di Bronzo* was awarded to *s.ten.* Ambrosio, Cattoi, Scaroni and *ten.* Ruffo di Calabria.

In January the unit fitted the first wireless sets on its aircraft, replacing the earlier forms of signals to the artillery, using firecrackers or lamps. Activity was limited in March, again because of bad weather, with just four missions, but it picked up in April. Overall, in the first four months of the year, it performed 76 war missions. On 15 April 1916 the 4ª Artiglieria became 44ª Squadriglia. Its commander, *magg.* Costanzi, became commander of the VII Gruppo.

5ª Squadriglia per l'Artiglieria

Above: Edoardo Barbiroglio, later a pilot with the 112ª Squadriglia, in training with a Gabardini powered by a tiny Anzani engine.

Formed on 15 October 1915 at Oleis on Caudron G.3, the 5ª Squadriglia was activated on 21 October and operated for the artillery command of the VIII Corpo d'Armata of the 2ª Armata. *Cap.* Raul Lampugnani was the commander with pilots *s.ten.* Cantoni and Giuseppe Retinò, *cpr.* Nello Barbieri and Luigi Panizzon, *soldati* Alberigo Balbi, Edoardo Barbiroglio and observers *sottotenenti* Francesco Allegra, Dario Armanino, Fabio Fabi and Raniero Sabatucci. Soon more observers arrived: *cap.* Vittorio Giovine, *ten.* Giuseppe Castaldi and *s.ten.* Appio Monti and Giulio Cesare Reggiani, and *soldato pilota* Luigi Fanti.

On 5 November 1915 the unit moved temporarily to Leproso airfield, near Oleis. During 1915 it performed 32 war missions.

On 1 January 1916 the pilots were *cap.* Lampugnani, *s.ten.* Cantoni and Giorgio Zoli, *serg.* Giuseppe Sarti, *cpr.* Barbiroglio and *sold.* Napoleone Rapini and the observers were *cap.* Giovine, *ten.* Monti, s.*tenenti* Allegra, Armanino and Reggiani. Commands praised the excellent missions of this unit in January. On 4 March *cap.* Lampugnani went to command the 1ª Squadriglia Artiglieria and *cap.* Vittorio Giovine replaced him. In March and April this squadron flew very little because of bad weather: only five war flights in the month of April. The 5ª Artiglieria became 45ª Squadriglia on 15 April 1916.

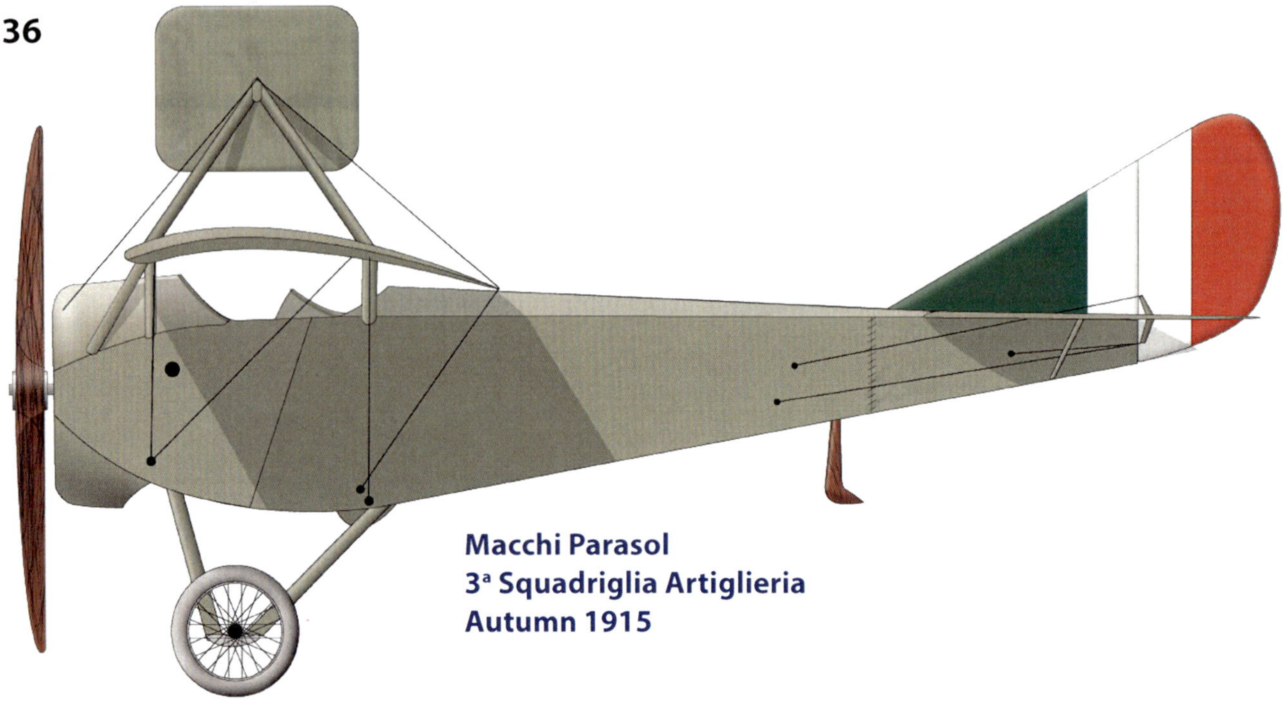

Macchi Parasol
3ª Squadriglia Artiglieria
Autumn 1915

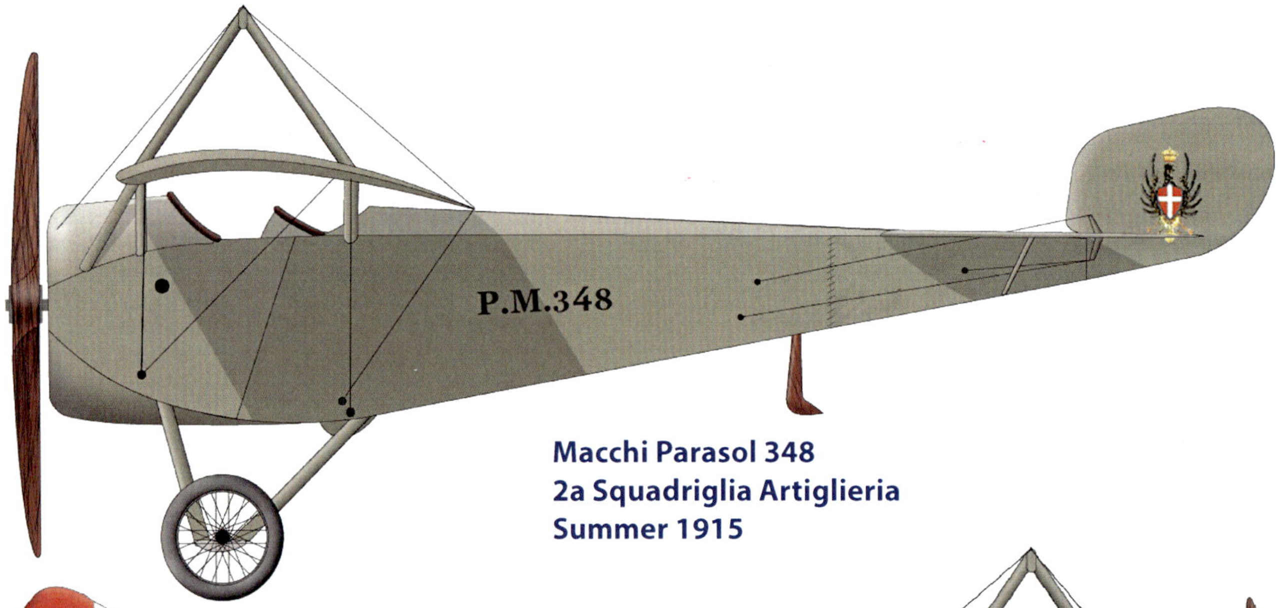

Macchi Parasol 348
2a Squadriglia Artiglieria
Summer 1915

Macchi Parasol
2a Squadriglia Artiglieria
Autumn 1915

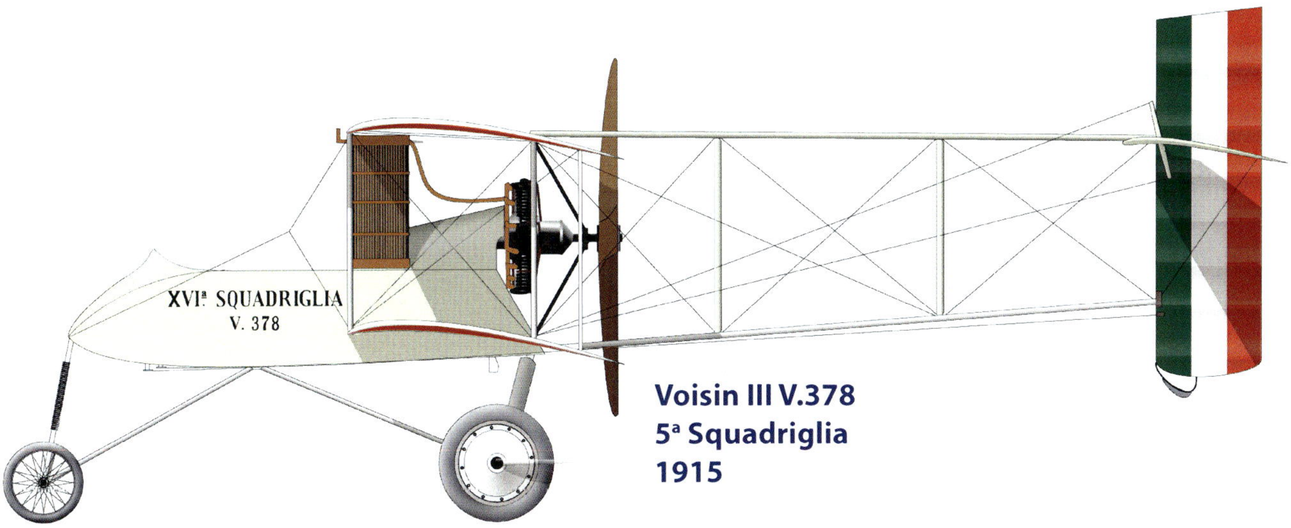

Voisin III V.378
5ª Squadriglia
1915

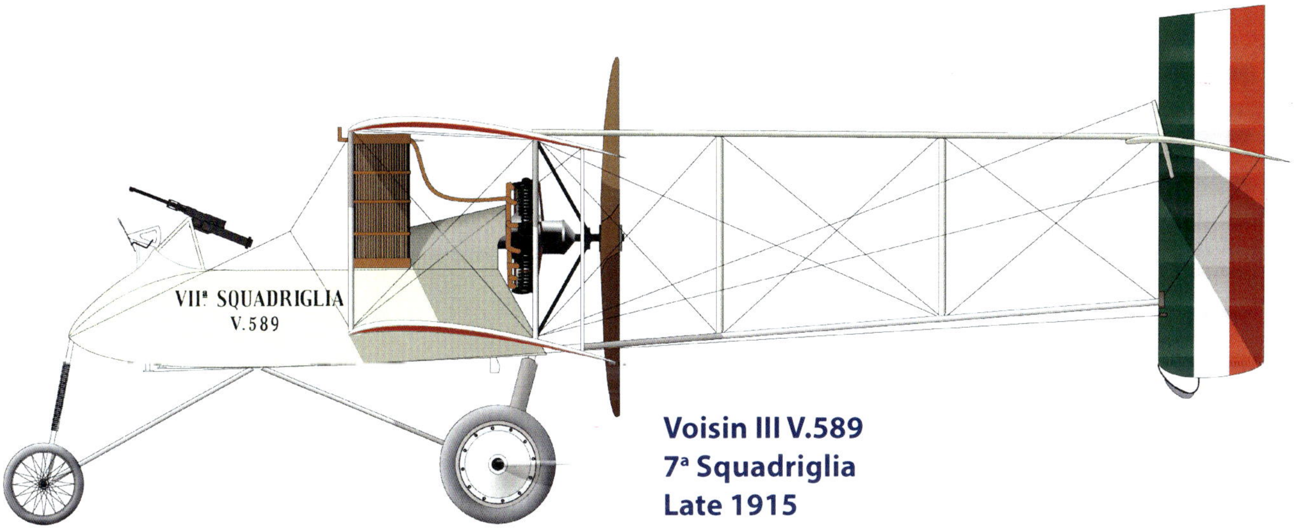

Voisin III V.589
7ª Squadriglia
Late 1915

Voisin III V.608
7ª Squadriglia
1916

6ª Squadriglia per l'Artiglieria

Formation of the 6ª Artiglieria began on 5 March 1916, on the Farman Fiat, and it was activated on 28 March at Pordenone with six pilots, two of them officers, none of them with the military brevet, so commander *cap.* Riccardo Adorni had to have them trained there. On 30 March 1916 *caporali* Antonio Corà and Lorenzo Serra during an alarm flight crashed to their death on Farman 756. Four expert observers arrived from other units, and four more from the observers' school at Centocelle. Its pilots included *s.ten.* Guido Conti, Pietro Cavadini, Ugo Lichtenberger, *cpr.* Vittorio Rial, *sold.* Vittorio Bosio; observers *s.ten.* Costantino Cattoi and *ten.* Mario Contaldi and Ugo Filangeri.

It went to Verona on 3 April with four Farmans, assigned to the northern 1ª Armata, flying its first mission on that day. On 13 April it detached some of its Farmans to Asiago. The 6ª became 46ª Squadriglia on 15 April 1916.

7ª Squadriglia per l'Artiglieria

The 7ª Squadriglia Artiglieria was born on 28 March 1916 from the re-assignment of the 9ª Farman. At the command of *cap.* Luigi Chiappelli, it was based at Oleis operating for the 2ª Armata, attached to the Gruppo Squadriglie per l'Artiglieria with a force of ten pilots, five observers and eight Farman Fiat. It became 47ª Squadriglia on 15 April 1916.

8ª Squadriglia per l'Artiglieria

The 8ª Squadriglia was formed at Turin at the end of March 1916, its commander was *cap.* Goffredo Maurel. It had pilots *s.ten.* Vittorio Bonomi and Rosario Bonelli, *cpr.* Sante Garofalo and Luigi Mancini, *soldati* Fedele Azzari and Manlio Pilati, Angelo Filippi and Raffaele Capo. It moved to La Comina and on 15 April 1916 it became 48ª Squadriglia.

Squadriglie da Caccia (Fighter Squadrons)

1ª Squadriglia Caccia: see 70ª Squadriglia
2ª Squadriglia Caccia: see 71ª Squadriglia
3ª Squadriglia Caccia: see 72ª Squadriglia
4ª Squadriglia Caccia: see 73ª Squadriglia
5ª Squadriglia Caccia: see 74ª Squadriglia

Squadriglie da Offesa (Bomber Squadrons)

The events of the Squadriglie da Offesa 1ª, 2ª, 3ª, 4ª, 5ª, 6ª, 7ª and 8ª appear in Part 5 of this book.

Part 5: The Bomber Squadrons from April 1916

1ª Squadriglia

Above: In front of a Caproni, the airmen of the first heavy bomber unit in the world. From the left, Laureati, Pellegrino, Carnevale, Graziani, Ercole, Barbieri, Piccio.

The 1ª Squadriglia da Offesa was formed on La Comina airport, near Pordenone, on 19 August 1915, initially just as 1ª Sezione Autonoma Ca 300 (Autonomous Ca 300 Flight) directly reporting to the Supreme Command. Its commander was *cap. pilota* Luigi Bailo, with pilots *cap.* Carlo Graziani, *tenente* Ercole Ercole, *sottotenente* Giulio Laureati and observers *magg.* Alfredo Barbieri, *cap.* Giacomo Agosti and Pico Deodato Cavalieri. From Malpensa the unit got Capronis 478 on 23 July, 480 on 1 August and 479 on 8 August.

After many test flights, the first war mission was on 20 August, a raid against the Austrian airfield at Aisovizza, (Haidenschaft) near Gorizia. Capronis 478, pilots Bailo and Graziani, and 480, Ercole and Laureati, each dropped six 162-mm bombs, six 135-mm bombs and incendiary bombs stored in the fuselage. On the way back, they strafed a Drachen, which was winched down. This action was reported in the War Bulletin, as a reprisal in the afternoon Austrian airplanes bombed Udine, killing five civilians.

The next day the two Capronis repeated the same action; in the afternoon General Cadorna, the Italian supreme commander, Gen. Porro, Vice chief of Staff and Gen. Diaz visited the unit and were flown on Ca 478 with *cap.* Bailo and Graziani as pilots.

On the 28th the same aircraft attacked again Aisovizza and this raid also was reported in the Daily Bulletin. On 1 September 1915 Ca 478 with *cap.* Bailo and Graziani flew for a test of the new Mod. E bombing sight. The next day

Above: Caproni 478, the first one in service, drew a lot of attention.

Ercole and Laureati flew on Ca 480 with three passengers, *cap.* Resio, Coniglio and *ten.* Bosio. On 4 September King Victor Emmanuel also came on a visit. Ercole and Laureati flew in front of the King. On 6 September an enemy airplane was spotted, two Caproni took off for a fruitless interception, and "air police" missions were flown also in the following days. On 8 Sept. Ca 478 and 480 flew to bomb the railway at Klause, but due to the cloud cover Ca 478 did not drop its bombs, and came back still carrying them, while Ca 480 of Ercole and Laureati with Barbieri and *cap.* Agosti completed the mission.

On 11 September Ca 478 attacked Monfalcone, but aborted due to mission for bad weather and technical problems; young crown prince Umberto, with his chaperons, came on a visit. On 12 Sept. Ca 478 and 480 flew reconnaissance missions over the frontline.

The big bombers were a magnet for visiting VIPs and foreign military observers. In early October the squadron was not tasked with any mission but it flew daily training flights dropping bombs, occasionally flying its Capronis in fruitless interceptions of enemy planes.

The squadron was active during the 3rd Battle of the Isonzo, fought from 18 October to 4 November. On 19 October *cap.* Bailo, Graziani and Barbieri with 478, Ercole, Laureati and Agosti with 479 attacked again Aisovizza, then one hour later the latter crew went out again with Ca 480 together with *cap.* Ernesto Pellegrino, *s.ten.* Ettore Marazzi and gunner Emilio Saletta, with their Ca 702, from 3ª Squadriglia which, both crew and aircraft, joined the 1ª Squadriglia. The unit attacked enemy installations all the following days, on 22 October Ca 480 of Bailo, Graziani and Barbieri was hit, the next day they flew Ca 479 which also was hit.

On 24 October Ca 702 met an enemy airplane, attacked it but after a few rounds the machine gun detached itself from its mount and fell down. The official Bulletin claimed an Aviatik driven away.

On the 29th again Ca 702 of Pellegrino and Marazzi attacked an Aviatik and drove it away, this event was once more reported in the bulletin of the Supreme Command. On November two more pilots arrived, *cap.* Ettore Croce and *cap.* Pietro Gregorini. Aisovizza was attacked again on 20 and 23 November.

On 30 November *cap.* Graziani left the unit to attend a course for staff officers and *cap.* Oreste Salomone replaced him, flying for the first time on 7 December. Bad weather precluded flights in December, on the 21st Ca 480 with Bailo, Laureati and Barbieri flew to Fiume for a long-range reconnaissance mission. In that period the 1ª Squadriglia had gunners Oreste Castoldi, Oreste Cornalba and Tonso and aircraft Ca 478, 479, 480 and 702.

The first mission of 1916 was a reconnaissance on 7 January that was not completed. On 13 January the 1ª Squadriglia was tasked to bomb Ljubljana, Ca 478 and 480 took off, but had to return on account of bad weather. The next day Ca 478 of Bailo and Salomone tried again but bombed Aisovizza. In the following days the Capronis were tasked with defensive patrols over Udine.

On 20 January 1916 all Capronis were grounded after an accident to Ca 708 of another unit, and an investigation found faulty strut attachment points, so all the aircraft were modified. Flights were resumed on February 5 with a mass flight of all Caproni squadrons to test close formation flying. These tests were repeated the following days.

On 14 February with a daring raid, Austrian aircraft bombed Milan, the second city of Italy, and a reprisal was soon under way. On 18 February a mass attack was launched against Laibach (Ljubljana) the capital of Slovenia. Caproni 478 "Aquila Romana" (Roman eagle) with Salomone, *cap.* Bailo and *t.col.* Barbieri and Ca 479 with Ercole and Laureati took off. The poet d'Annunzio was supposed to be on Ca 478, but as he was late at La Comina that morning, *t.col.* Barbieri, commander of the Battaglione Squadriglie Aviatori, the Italian air service, insisted on taking his place and flew to his death. Ca 478 faced the attack of two Fokker A.III (Austrian designation of the E.1) of pilots *Hauptmann* Heinrich Kostrba and *Hauptmann* Mathias Bernath. *Cap.* Bailo and Barbieri manning the front machine gun were killed and a bullet grazed Salomone's head. Salomone brought back the Caproni, still loaded with bombs, to a landing at Gonars airfield. Ca 479 completed the mission with an engine stopped, landing safely at la Comina, while 4ª Squadriglia lost Caproni 703. Three days later on his own initiative the King assigned the *Medaglia d'Oro al Valor Militare* (gold medal for military valour, Italy's highest decoration) to Salomone, the first one earned by an aviator,

Above & Below: The commander in chief of the Italian army, Gen. Cadorna, and his deputy, Gen. Porro, on 21 August 1915 flew on Ca 478 for 25 minutes with pilots Bailo and Graziani.

Above: On 18 February 1916 Salomone brought back Ca 478 with its nacelle stained with blood.

and commander in chief Gen. Cadorna himself pinned it on his breast.

As Salomone recovered, *cap.* Ercole was acting commander of the 1ª Squadriglia. On 4 March 1916 the unit was assigned two new pilots, *s.ten.* Edoardo Scavini, who remained just a few days, and *serg.* Giuseppe Martinetti.

The demonstration of the capacity of fighters with synchronized guns had left its mark, and a device to fit a machine gun in the back of the Caproni, designed by Laureati, was tested on Ca 479 on 16 March, with *magg.* Abiel Bethel Revelli and technicians from Turin attending. The unit returned to action on 2 April, Ca 479 and 480 bombing Adelsberg and the new gun was successfully used driving off a Fokker that had placed itself on the tail of a Caproni. New pilots arrived, *s.ten.* Cola De Renzis, *brig.* Annibale Comazzi and *sold. vol.* Domenico De Lorenzi.

On 19 April *cap.* Salomone returned to his command, the unit meanwhile having taken the name of 1ª Squadriglia Caproni and the Gruppo Comando Battaglione Aviatori had become IV Gruppo. After two weeks of bad weather, the unit was back in action on 20 April, Ca 480 with *s.ten.* De Renzis, *serg.* Martinetti and gunners Cornalba and Emidio Bonora bombed the seaplane station of Trieste. On 25 April the unit got Ca 300 hp 1139. On 26 April Ca 479 with the crew De Renzis, Martinetti, Cornalba, Olivieri bombed Ovcia Drava. *Cap.* Ercole went to command a new squadron for Albania.

On 14 May the Austrian Chief of Staff Conrad launched the Stafexpedition (punishing expedition) against Italian defences in the area of the "altipiani", plateaus, between Vicenza and Trento. This offensive at aimed at cutting through the Venetian plain, behind the bulk of the Italian army facing east along the Isonzo River. The Supreme Command used its Caproni squadrons to bomb the positions of the attackers. The 1ª Squadriglia bombed camps in the woods of Campolongo on 25 May with Ca 479 (crew De Renzis, Comazzi, gunner sold. Joren Napoli) using incendiary bombs, Roana on May 30 with Ca 1139 (De Renzis and Martinetti), in Val d'Astico. Other actions on the Asiago front were on June 2 (Ca 1139 De Renzis/Comazzi, on the way back the plane force landed near Conegliano, but the problem was fixed and it returned to base), on June 7 near Mount Verena (Ca 1139 De Renzis/Martinetti) and

Above: Gen. Paolo Morrone, Minister of War from April 1916 to June 1917, is shown the Capronis of 1ª Squadriglia. Behind him are *magg*. La Polla, commander of the bomber group, and *cap*. Mondelli.

Right: The crash of Caproni 1257 of *ten*. Errera and *serg*. Fossati. They crashed on 1 May 1917.

on 9 June in Val d'Astico (Ca 1139 Laureati/Comazzi). On June 11 the target of Ca 1139, with pilots De Lorenzi and Martinetti was Camporovere, on 16 June (Laureati and Comazzi) Campomulo again. Large quantities of steel darts were also dropped on infantry in the field. On 24 June De Renzis and Comazzi attacked a 305-mm gun at Campomulo, on 26 June the same pilots bombed encampments at Rotzo.

On 1 July the unit, that had a new pilot, *cpr.* Giovanni Castigliano, got Caproni 1181, with central 150 hp Isotta Fraschini engine, commonly called 350 hp. It took off on its first mission on 4 July (Laureati and Martinelli) but it aborted on account of strong wind and a failed engine. The mission was completed on July 6 with the same crew. In the mission of July 15 Ca 1139 of De Lorenzi and Castigliano had its wings badly damaged by flak. Old Ca 479, after one year of service, was sent back to Milan, and Laureati returned with the new Ca 1233, a 300 hp, on July 22.

On 29 July the 1ª Squadriglia got Ca 1247, another 350 hp. The plane came in from Taliedo with pilots *ten.* Marazzi and *sold.* Dall'Oca and then was exhibited at La Comina for Prime Minister Paolo Boselli.

On 1 August three Capronis were prepared for a large operation, a raid against the Danubius works and the torpedo factory at Fiume (now Rijeka, Croatia), but a series of problems stopped the take-off of Ca 1139 (De Renzis/ Comazzi) four times, while Ca 1233 (De Lorenzi/Castigliano) lost contact with the other ones and aborted the mission, and only Ca 1181 (Laureati/Martinetti) with *magg.* Alberto Marenco and *s.ten.* Guzzanti as gunner, reached its target.

In the first two weeks of August 1916 the 1ª Squadriglia supported the 6th Battle of the Isonzo, that brought the conquest of Gorizia, and attacked railroads at Opcina, Dornberg and Prvacina. On 9 August a Fokker and a Brandenburg attacked Ca 1247 but were repulsed.

On 25 August the 1ª Squadriglia had Capronis 1139, 1181, 1233, 1247 and 1232. Its force included commander Salomone, pilots *cap.* Laureati, *ten.* De Renzis, *brig.* Comazzi, *serg.m.* Ettore Borra, *serg.* Martinetti, *caporali* Castigliano and Alessandro Fossati; observers *cap.* Tullio Toccolini, *s.ten.* Luigi Guzzanti and Giambattista Pratesi; gunners Castoldi, Arrigo Andri, Niccolò Joren Napoli and Pio Serafini. Pratesi was the cover name of Della Martina, an *irredento*, that is to say an Italian with Austrian citizenship. On 5 Sept. *s.ten pil.* Giovanni Zerpp arirved. Night training flights began at Aviano with the aid of three searchlights.

On 13 September Salomone returned to action, flying Ca 1247 with De Renzis, Castoldi, Andri, together with 1181, (Laureati, Borra, Guzzanti) and 1233 (Comazzi, Castigliano, Pratesi) bombing the Lloyd yards at Trieste with Nieuports as escort.

On 17 September Ca 1233 (Zerpp, Castigliano, Pratesi and gunner Pio Serafini) and Ca 1232 (De Renzis, Fossati, Andri, Joren) bombed the railroad at Dottogliano In the night of 25 September Laureati flew five times with Ca 1139 to train the other pilots. Then for one month there were no combat flights. On 17 October a French military mission formed by Maj. Faure and Capt. Le Prieur visited the squadron with *gen.* Marieni, commander of the Italian Air Force. On 28 October *s.ten. pil.* Ottorino Manzoni arrived. Only on 31 October the unit returned to action, bombing Dottogliano with Ca 1233 (Zerpp, Castigliano, Pratesi, Castoldi) and Ca 1247 (De Renzis, Fossati, Guzzanti, Serafini). On 1 November the target was encampments at Cernizza for Ca 1181 of Comazzi, Borra, Guzzanti, Joren. In November Laureati and Salomone were assigned to Commands in Turin and *ten.* De Renzis became acting commander. On 26 November Ca 1257 300 hp arrived.

After another month with no action, on 3 December two aircraft were tasked to bomb the station at Dottogliano. *Linienschiffsleutnant* Gottfried Banfield, flying from Trieste on seaplane A12 attacked Ca 1233 with pilots *s.ten.* Manzoni and *serg.m.* Borra with the observer *s.ten* Guzzanti and gunner Castoldi, near Sesana, together with *Zgsf.* Karl Cislaghi of Flik 28 and Godwin Brumowski on Brandenburg D.1 65.53. Gunner Castoldi was killed, the Caproni crashed near Mavinje and the survivors were captured. No more war missions were flown in 1916.

In 1917 commander of the 1ª Squadriglia was *cap.* Toccolini with pilots *ten.* De Renzis, *s.ten.* Giovanni Zerpp and Gilberto Errera, Comazzi, now *aspirante*, *cpr.* Castigliano and Fossati; observers *s.ten.* Pratesi and *ten.* Mario Venturini; gunners Andri, Joren Napoli, Serafini, Carlo Barberis, Angelo Broggini, Girola and Libona. Its equipment consisted of Caproni 1139, 1181, 1247 and 1257.

In the night of Jan. 11 Ca 1181 (Zerpp, Castigliano, Pratesi, Andri) bombed the Lloyd yards at Trieste and the seaplane base at Prosecco. A new pilot arrived, *s.ten.* Guido Taramelli and on January 20 *cap.* Domenico Mondelli became the new commander.

From 25 January to 8 March 1917 no flights were made. On 9 March training flights in formation began. The 1ª Squadriglia didn't fly any war mission on account of bad weather until 13 April, when it returned to action attacking Prvacina with Caproni 1247 (Taramelli, Venturini, Castigliano, Andri) and 1257 (Errera, Fossati, Joren, Broggini) that suffered a failed central engine and bombed the alternative target, Ovcia Draga station. On 20 April Mondelli and Comazzi and gunners Joren Napoli and Ugo Campi bombed Chiapovano on Ca 1181. A ball trailing a steel chain wounded Comazzi in the face and fell in the cockpit, the aircraft landed at S. Maria La Longa. The other bomber on that mission, Caproni 1257 of Fossati and Errera, bombed Idria and faced no less than ten attacks by Brandeburg C.1s but gunners Broggini and Andri kept them at bay. In his back turret, Andri had his fingers frost-bitten.

On 1 May 1917 two bombers flew out in the night to attack Opcina and got lost in the fog. Ca 1181 of *cpr.* Castigliano and *ten.* Guido Taramelli landed on the banks of the Piave River still with all of its bombs, (the plane was recovered), but Ca 1257 of *serg.* Fossati, *ten.* Errera and Broggini which did bomb its target also force landed on the Piave but was wrecked, injuring Fossati. The official Bulletin

Above & Below: A Caproni Ca.3 of 1ª Squadriglia at San Pelagio in 1918. Two Fiat machine guns are on the back turret.

just said: "*Two of our bombers bombed last night the railroad station at Opcina and then returned safely to their field.*"

In this period 1ª Squadriglia had pilots *cap.* Mondelli, *ten.* De Renzis, Taramelli, Zerpp, *s.ten.* Errera, *asp.* Comazzi, *serg.* Fossati, Castigliano, G. Battista Trotti, observers *cap.* Ernesto Fogolà, M. Aurelio Barbarisi, Toccolini, s.ten. Pratesi and Venturini, gunners *soldati* Nicolò Joren Napoli, Andri, Serafini, Broggini, Campi, Emilio Baroni, Bonora and Agostino Montaldo.

In May 1917, supporting the 10th Battle of the Isonzo, the 1ª Squadriglia was in action: on 14 May Ca 1247 (Zerpp, Trotti, Barbarisi, Joren) bombed Vojensko and dueled with an Austrian plane, Ca 1190 (Mondelli, Comazzi, Broggini, Bonora) aborted the mission. On 18 May the same two planes bombed Canale, on 23 May three bombers, 1166, 1190 and 1247 bombed trenches at Vojscica, enemy fighters hit 1166 of Mondelli and Castigliano. On 23 May 1917 *s.ten.* Zerpp was removed, and assigned to an artillery unit, because he had bombed Italian trenches by mistake. Mondelli's air career also ended, but the valiant officer was included in the staff of Gen. Capello, commander of the 2ª Armata, during the 11th Battle he was the only Italian officer that entered Castagnevizza, and he was called "*il negro meraviglioso*", the wonderful Negro. Two planes repeated the same action the next day, when no less than 130 aircraft where in action.

In the first week of June the pilots flew their old Caproni back to Taliedo and returned with new Ca.3 450 hp 2385. 2386, 2389 and 2390. They went into action on 10 June, supporting the Battle of Ortigara, an Italian attack in the north. Ca 2386 and 2389, with pilots Taramelli, Errera, Comazzi and Castigliano flew to Campo Gallina but did not drop the bombs due to the clouds. On 15 June Ca 2386 (Taramelli, Errera, Joren, Montaldo) 2389 (Castigliano, *cpr.* Antonio Bresil, Venturini, Andri) and 2385 (Comazzi, *sold.* Antonio Cominotti, Fogolà, Bonora) bombed effectively the station at Baca di Tolmino, accurate ground fire hit their planes. On 19 June, a day in which 145 Italian planes were in action, the same Caproni and crews bombed Campo Gallina, Austrian fighters attacked and damaged two of them, but escorting fighters drove them away. In the night of 26 June Ca 2385 (Taramelli, Errera, Venturini, Broggini) bombed targets at Nabresina and Prosecco, at the south-east limit of the front, following the route marked by searchlights. The Bulletin reported that a whole Squadriglia had bombed. The same crew and plane flew a similar mission on 29 June but couldn't complete it. On 30 June three bombers with the usual crews attacked Mavlinje in daytime *serg. pil.* Manfredo Benini flew his first war mission.

On 7 July a group of Caproni, among them 2386 and 2389, with good fighter escort attacked the power stations and the mercury refining plant at Idria, important for the production of explosives.

On 23 July the target was San Daniele del Carso but the formation bombed targets of convenience due to bad weather. Three Capronis of 1ª Squadriglia flew, 2386, 2389 and 2390 carrying *magg.* Falchi. On 28 July the target was again the plant at Idria where 4 tons of bombs were dropped. The escorting fighters shot down two enemies: 2386, 2389 and 2390 with the usual crews flew in tight formation with the other bombers. *Magg.* Falchi, piloting 2390, was wounded by ground fire.

Only two planes, 2389 and 2386 took part in the night bombing of Pola on 2 August. Both reported accuracy in their bombing and both were hit by the accurate ground fire. The mission to Pola was repeated the next night, 2386, 2389 and 2390 with the usual crews and a new observer, *ten.* Ugo Giuffrè, 2389 and 2390 had to turn back due to engine troubles, 2386 (Taramelli, Errera, Venturini, Broggini) hit the arsenal.

On 7 August the target was again Chiapovano for 2385 and 2386, 2390 of ten. Marazzi introducing to combat a new pilot, *cpr.* Giuseppe Modena, had engines trouble and dropped its bombs in the countryside.

Pola was attacked again in the night 7/8 August 1917 by 2385 and 2389 and the new Ca3 4041 (Marazzi, Comazzi, Fogolà, Joren). Having dropped its bombs, Ca 2389 of Castigliano, Bresil, Giuffrè and gunner Bonora faced two attacks from an Austrian seaplane over the sea, but returned to base. The next missions were an attack on Chiapovano, 11 August, and on the iron works at Javornik, 14 August, for four bombers, 2389. 4041, 2385 and 2386 of *ten.* Taramelli, who received a *Medaglia d'Argento*, for having flown at low level to strafe the enemy.

In mid-August 1917 the 1ª Squadriglia at La Comina had Caproni 450 hp serials 2385, 2386, 2389, 2390 and 4041, marked with red stars. With its commander *cap.* Ernesto Fogolà there were pilots *ten.* Taramelli, *s.ten.* Comazzi, Errera and Marazzi, *asp.* Castigliano, *cpr.* Bresil and Modena; observers *ten.* Venturini and Giuffrè; gunners Andri, Bonora, Broggini, Joren Napoli, Montaldo and Giorgio Sacco.

On 17 August the Italians launched the biggest offensive of the war, the 11th Battle of the Isonzo, that lasted until 12 September. The 2nd Army forced the river but the attack, that conquered the Bainsizza plateau and placed the *Isonzo Armee* of Gen. Boroevic on the brink of collapse, wore itself out and failed to conquer Tolmino and Mount S. Gabriele. The air force took part in the battle with impressive intensity, and a huge use of force. The 1ª Squadriglia was in action on 18 August, three fighters attacked Caproni 2386 of *ten.* Taramelli and *serg.* Emilio Raimondo. Gunner Broggini hit a Brandenburg KD fighter and saw it leaving the combat out of control. Fifty Austrian bullets were then found in the Caproni. Ca 2389 and 4041 were also hit.

The next day, 19 August, the unit went into action twice with three planes. An Austrian fighter with a black striped tail attacked again 2386, now with pilots Taramelli and Errera. Immediately a Spad and a Nieuport engaged the fighter. On that day *ten.* Marazzani of the 77ª Squadriglia claimed his first victory. On August 20 once again three Capronis bombed the same targets, shooting at batteries on the Hermada plateau with their machine guns. *Asp.* Castigliano was awarded a Silver Medal. Three Caproni were back in action on 21 August, and this time *ten.* Errera was awarded the Silver Medal for daring low level attack. In the afternoon a new mission was ordered to Ivanigrad, two Caproni took part, 2389 (Comazzi, Castigliano, Bonora and *ten.* Giacomo Macchi, an observer of 25ª Squadriglia) and 2386 (Taramelli, Errera, Venturini, Biagini) dropping incendiary bombs attached to parachutes, which proved useless. The Capronis of 1ª Squadriglia were again in action on 22 and 23 August, but on the 24th out of three bombers attacking Hudajuzna one failed to return, Ca 2389 of Comazzi, Benini, Fogolà and Joren Napoli, victim of anti-aircraft guns at S. Lucia di Tolmino. There were no survivors. Strangely, the official diary of 1ª Squadriglia does not report this loss.

On 26 August d'Annunzio flew on Caproni 2378 with pilots *cap.* Maurizio Pagliano and *ten.* Luigi Gori, from 8ª with the observer Pratesi. Together with Ca 4041and 2386 they bombed Lokva. On 28 August two planes bombed the station of Grahovo, then again on 30 August.

On 1 September Ca 2386 and 2390 with the new pilots *m.llo* Salvatore Sambataro and *cpr.* Pietro Piaser went in action over Mount San Gabriele.

In the night of 3 September three Capronis 2386, 4041 and 2408 (pilots Errera and Sambataro with GM Mario Savino as observer) attacked the naval base of Pola, 2386 was badly shot up by an Austrian seaplane. The long serving *ten.* Marazzi gained a Silver Medal.

Pairs of Capronis flew missions on 4, 5 and 6 September against Ternova and Lokve, then there were no more combats until 15 September, when four Capronis attacked and again on 24 September over Doberdò. After sunset on 25 September there was an attack to the station of Prosecco by 2407, 2378, 2386, and 4041 with two new observers *s.ten.* Marcantonio Ferrucci and *ten.* Pietro Scotti Fantoni. On 27 September three Capronis attacked again the railroad at Doberdò, 2387 of Marazzi, but Modena had to give up with engine problems.

A new night mission against Pola in the night of 28 September, for 4041 with *cap.* Darby and *ten. oss.* Rasi, from 8ª Sq., which suffered the attack of two seaplanes, and 2407. The next day, 29 September, mission to Comen for 2407 of Taramelli and Raimondo and in the night Pola again, the planes were 2386 and the new Ca 4171 of Castigliano and Bresil that failed to reach the target and bombed a secondary target

On 1 October the 1ª Squadriglia of the IV Gruppo at La Comina had Capronis 2386, 2407, 4041, 4145 and 4171. Its temporary commander was *ten.* Marazzi, and its force included pilots *ten.* Errera and Taramelli, *m.llo capo* Sambataro and *serg.* Bresil, Castigliano, Raimondo and Pietro Piaser, *cpr.* Modena; observers Mario Venturini, *asp. GM* Augusto Anfossi, *ten.* Pietro Scotti Fantoni and *s.ten.* Marcantonio Ferrucci; gunners Andri, Broggini, Bonora, Campi, Montaldo and Giuseppe Mora.

On 2 October Castigliano, Bresil, Anfossi and Andri on 4171 attacked the station of Grahovo. In the night, Ca 4145, Errera and Taramelli, bombed Pola again, then no further actions are recorded.

On 24 October the Austrians and Germans launched operation "Waffentreu" breaking through the Italian front at Plezzo and along the Isonzo River to Caporetto.

The squadron was heavily engaged trying to contain the threatening advance of the enemy troops. Two Capronis, 2328 with the commander of IV Gruppo *magg.* Bonazzi and 4171 bombed Caporetto on 25 October.

On 26 October the unit repeatedly bombed S. Lucia di Tolmino, each time clashing with enemy fighters: Ca 4171 of Castigliano and Raimondo, twice attacked by enemy fighters which put one engine out of action, 4145, Taramelli and Piaser, 2386 with Castigliano and Raimondo, completed their second sortie that day with just two engines running. And at night, 4145, Taramelli and Sambataro and 4190, Marazzi and Modena.

On 30 October 4041 of *magg.* Bonazzi and *ten.* Giampiero Clerici and gunner Bonora carrying *col.* Riccardo Moizo, the air force commander, with 2328, Errera and Piaser attacked Ponte Torre, then 4171, Taramelli and Raimondo with gunner *cpr.m.* Luigi Bellini and 2328, Righi and Modena, bombed it again.

On 31 October 4171 and 2328 bombed Osoppo, then 4171 flew again with just pilot Castigliano and two gunners to drop bread to Italian troops near Codroipo.

On that day the 1ª Squadriglia was ordered to retreat from La Comina to San Pelagio, near Padua. Ca 4171 took off with pilot Castigliano and two gunners, but it was wrecked near Treviso; 2386 with Sambataro and three mechanics, 4041 with Piaser and *s.ten.* Molteni, 4190 with Errera, Bresil and three mechanics, 2328 with *s.ten.* Giampiero Clerici and Raimondo landed at San Pelagio after having dropped bread, 2407 with Taramelli, Ferrucci and three mechanics and finally 4145 with *asp.* Stefano Righi and Modena arrived *Cap.* Pagliano was ordered to stay at la Comina to destroy buildings and equipment before the arrival of the enemy, which he did on 5 November.

On 1 November Caproni 4190 flew to La Comina to pick up the last pilots. Caproni 2338 with *s.ten.* Clerici and *ten.* Domenico De Lorenzi, observer *ten.* Ignazio Lanza di Trabia, of 25ª Sq., a Sicilian prince, and gunner Cosimo Pisanello failed to return from a mission to Udine. Nobody survived, possibly victims of *Vizefeldwebel* Ueberschär of Jasta 39. Clerici's brother, Umberto, also a Caproni pilot of 2ª Squadriglia, had died in 1916.

In the following days the unit was very active, bombing by day and night bridges and roads. Sorties were flown on 3, 4, 5, and 6 November. On 13 November the target was Zenson, at the Piave River, for Ca 4145 and 4190, where ground fire wounded *s.ten.* Ferrucci. There were actions every day. On 15 November Ca 4145 carried *magg.* Bonazzi and *col.* Moizo. Their targets ranged from Primolano to Zenson, gunners Montaldo Daveri and Alessandro Ranuzzi had their baptism of fire. On 20 November four Capronis attacked the area of Zenson, the Austrian pocket across the Piave River.

Above: Two American pilots with the 1ª Squadriglia. Fuselage roundels have been introduced.

On 23 November four Capronis attacked the enemy airfield at Feltre and *cap.* Pagliano, now squadron commander, on 4145, flew with with *magg.* Bonazzi. Bombing missions were daily.

As the front stabilized along the Piave River, the final enemy threat came from the mountain junction from the Astico to the Grappa, and only in early December could the Italians contain it. On that sector of the front, the squadron attacked Galmarara with four bombers in this last attack. On 3 attacked targets at Gallio, Campomulo and Val Galmarara, in this last attack, on 3 December the temperature was -7.6° F, and some airmen were frostbitten. Many more missions followed, On 30 December the 1ª Caproni bombed the airfield at Codega di S. Urbano. One bomber failed to return, Caproni 4216 of Pagliano and Gori with gunners Giacomo Caglio and Arrigo Andri. Benno Fiala, of Flik 56/J, flying Albatros D.III 153.77 intercepted it over Susegana and shot it down in flames. There were no survivors. Aviano Air Base, home of the USAF 31st Fighter Wing, is named after Pagliano and Gori.

Taramelli, by now *capitano*, became the new commander. Not all data are available for the numerous actions by day and night in January 1918. In the moonless night of 2 February 1918 three aircraft, with pilots Errera and Sambataro, Bonazzi and Marazzi, Taramelli and *s.ten.* Vittorio Moruzzi went out to bomb Monte Rovere. On the way back Ca 4233 got lost in the fog and was damaged in a forced landing near Legnago. Pilot *s.ten.* Moruzzi was killed, Taramelli, Tubertini and Montaldo were injured. Other bombing actions followed, on 19 February the Caproni of *cap.* Bergonzi and Bonazzi again carried *col.* Moizo. In the night of 26/27 February the unit bombed the station of Bolzano after a flight of 300 km over the Alps. The pilots were Bonazzi and Errera, Castigliano and Modena. The action was repeated the next night.

On 16 March during a training flight Ca 4223 crashed due to an error of its inexperienced pilot *s.ten.* Giuseppe Federici, of the 13ª Squadriglia, killing *serg.* Giuseppe Modena and gunners Campi and Bonora, and a woman on the ground.

In May 1918 the 1ª Squadriglia was at San Pelagio with commander *cap.* Taramelli, equipped with Caproni 2337, 4145, 11499, 11503 and 11505. Its pilots then *ten.* Errera, Marazzi, Lelio Lucherini, *serg.* Bresil, Pietro Piaser, Adolfo Picchetta, Emilio Raimondo and Mario Rossi; observers *ten.* Venturini, Cesare Teofani and Giuseppe Tubertini and *s.ten.* Ferrucci, Angelo Copello and Arturo Roseti; gunners Barberis, Oreste Borriello, Mora, Riccardo Daveri, Angelo Di Fede, Gaetano Ferrara, Giuseppe Pretonari, Adelchi Ravagnani, and Luigi Soldà. Sorties went on against railroad yards and airfields. At this time Taramelli had flown 58 bombing missions, Errera 57.

During the Austrian offensive of mid-June, the Capronis attacked repeatedly S. Donà and the main roads, often carrying as observer Gabriellino d'Annunzio, son of the poet. On 16 June during an attack to a gangway across the Piave a shell killed Tubertini, his aeroplane, Caproni 11505 was slightly damaged but all other crew members, *serg.* Bresil and *magg.* Bonazzi, were unhurt. That day Ca 4145 force landed due to an engine failure and crashed, injuring Lucherini and observer Ferrucci, while Raimondo and Ferrara were unhurt.

In June 1918 the unit got American pilots trained at Foggia: Lieutenants Raymond P. Baldwin, Kenneth G. Collins, Norman Terry and Alfred S.R. Wilson and *s.ten.* Giorgio Cordero di Montezemolo.

In July sorties were flown dropping only leaflets.

On 1 August at their first mission, bombardment of the airfield at Motta di Livenza, Cordero di Montezemolo and Collins had their plane badly damaged, making an emergency landing at Marcon.

The former Italian airfields, now in enemy hands, were bombed by night in August. On 17 September Ca 11505 attacked from low altitude the airfield of Portobuffolé and bombed a gun battery that was firing at the attacking bombers, putting it out of action.

On 26 October Ca 4162 crashed during a test flight, killing *ten.* Mario Simonini, *s.ten.* Giorgio Cordero di Montezemolo and Antonio Menegazzi and fitter Bernardo Agazzi. The final Italian offensive had started, and the troops of General Diaz broke through at Vittorio Veneto. The squadron attacked repeatedly the enemy airfields during the preparation, and then went into action against the roads of the retreating Austrians, strafing from low level. The last combat mission of the war was flown in the afternoon of 30 October, when Caproni 11505 with pilots Bresil and *magg.* Fernando Bonazzi, with Daveri and *t.col.* Egidio Carta, bombed the road from Sacile to Pordenone and strafed retreating troops.

On 4 November the unit was at San Pelagio still with Lucherini as commander, with Caproni 2337, 11499 and 11505. Its pilots were Lieuts. Collins, Terry, Wilson, *s.ten.* Osvaldo Bertosi, *sergenti* Bresil, Piaser, Raimondo, Gaspare Milanesi and Pietro Mingozzi; its observers were *tenenti* Mario Pansa, Andrea Valfrè di Bonzo, *s.ten.* Ferrucci, Filippo Beccaria and Giuseppe Mezzano; gunners Barberis, Daveri, Di Fede, Soldà, Giambattista Carabelli, Daniele Damelli and Adelchi Ravagnani.

In 1919 the 1ª Squadriglia moved to Aspern, Austria, supporting the armistice commission, and its aircraft operated on detachments to Prague and Krakow. It returned to Italy at the end of the year and was disbanded.

Casualties and Combat Losses of the 1ª Squadriglia

Date	Place	Aircraft	Crew	Cause
18.2.1916	Merna	Ca 478	*Cap.* Salomone WIA, *cap.* Bailo KIA, *t.col.* Barbieri KIA	Fighters
3.12.1916	Selo-Jamiano	Ca 1233	*S.ten.* Manzoni POW, *s.m.* Borra POW, *s.ten.* Guzzanti POW, *cpr.m.* Castoldi KIA	Fighters
24.8.1917	S. Lucia	Ca 2389	*S.ten.* Comazzi KIA, *serg.* Benini KIA, *cap.* Fogolà KIA, *sold.* Joren Napoli KIA	Ground fire
1.11.1917	Flumignano	Ca 2328	*S.ten.* Clerici KIA, *ten.* De Lorenzi KIA, *ten.* Lanza KIA, *sold.* Pisanello KIA	Fighters
30.12.1917	Susegana	Ca 4216	*Cap.* Pagliano KIA, *ten.* Gori KIA, *sold.* Caglio KIA, *sold.* Andri KIA	Fighters
2.2.1918	Legnago	Ca 4233	*S.ten.* Moruzzi KIC, *cap.* Taramelli WIC, *Ten.* Tubertini WIC, *sold.* Montaldo WIC	Accident
16.3.1918	San Pelagio	Ca 4223	*S.ten.* Federici KIC, *serg.* Modena KIC, *sold.* Campi KIC, *sold.* Bonora KIC	Accident
16.6.1918	S. Donà	Ca 11505	*Ten.* Tubertini KIA, *magg.* Bonazzi UNH, *serg.* Bresil UNH	Ground fire
26.10.1918	San Pelagio	Ca 4162	*Ten.* Simonini KIC, *s.ten.* Cordero KIC, *s.ten.* Menegazzi KIC, Agazzi KIC	Accident

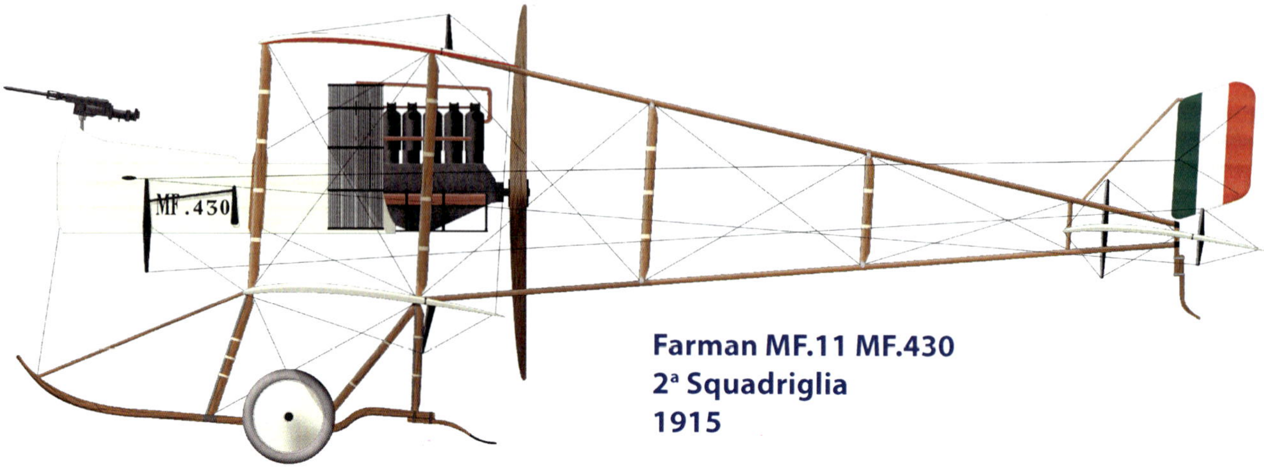

Farman MF.11 MF.430
2ª Squadriglia
1915

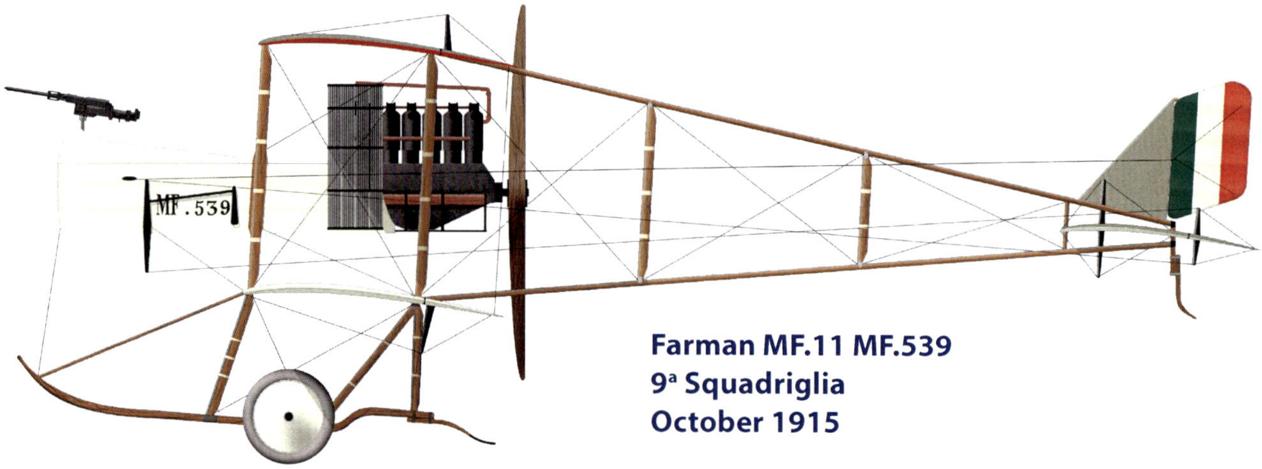

Farman MF.11 MF.539
9ª Squadriglia
October 1915

Farman MF.11 MF.268
9ª Squadriglia
Summer 1915

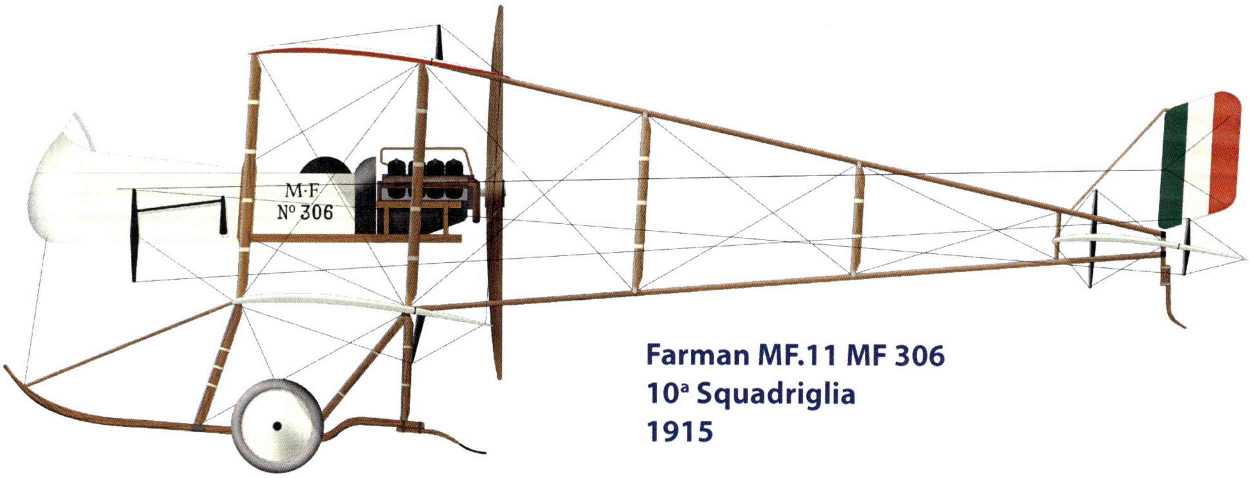

Farman MF.11 MF 306
10ª Squadriglia
1915

Caproni Ca.3
Unit unknown

Caproni Ca.3
Unit unknown

2ª Squadriglia

Above: Ca 483, the first bomber of 2ª Squadriglia, loaded with bombs.

This unit was formed at Pordenone on 17 October 1915 as 2ª Squadriglia Caproni with commander *cap. pilota* Egidio Carta. It was assigned to the Command of the Battaglione Squadriglie Aviatori, directly within the Supreme Command, and it was based at Aviano. Its original force was Caproni 300 hp serials 483 and 487 with one more as reserve, 479. It had pilots *cap.* Ettore Croce, *cap.* Pietro Gregorini and *serg.m.* Virgilio Sala with the observer *s.ten.* Gustavo Herman.

It went into action immediately during the 3rd Battle of the Isonzo, attacking on 19 October the enemy airfield of Aisovizza with two airplanes, repeating the action the next day. In this period, it used as observers *magg.* Gallina and Moizo and *cap.* Berliri Zoppi and Pico Deodato Cavalieri, *ten.* Fausto Pesci and *s.ten.* Speltini. It flew several missions, particularly against the Austrian airfield at Aisovizza.

On 1 December the unit took the new name of 2ª *Squadriglia da offesa*. Up to the end of the year it flew a total of 16 bombing missions for a total of 27 combat sorties, including both reconnaissance and bombing missions, participating in the 4th Battle of the Isonzo with attacks against barracks and communications. It received pilots *ten.* Umberto Clerici and Alberto Verona, *serg.* Virgilio Sala (later test pilot at Caproni and a General) and *cpr.* Oscar Lazzerini.

Their tasks were sometimes weird, like the mission on 9 January 1916: to ascertain if five new division were being moved to the front: the crew of *cap.* Carta, *serg.* Sala and *cap.* Guido Barutta could only spot some trains, and strafe a Drachen along the way. This was the only combat flight in the first trimester of 1916.

On 19 January 1916 the Supreme Command stopped the flights of the Caproni on account of weakness of the wing attachment bolts. Flights resumed at the end of January, but bad weather curtailed activities until April.

On 1 April the 2ª Squadriglia at Aviano with commander Carta had Capronis 483, 708, 712. Its force included pilots *s.ten* Sala, *ten.* Clerici, *ten.* Verona, *serg.* Ernesto Canelli and *serg.* Lazzarini; observer *ten.* Umberto Fioretti and gunners Massimo Chiaperotti and Emilio Saletta.

On 2 April there was the first combat sortie of the year, with Ca 483 of Clerici, Sala, Fioretti and Saletta participating in the bombing of Adelsberg. Austrian fighters intercepted

Above: Ca.3 2371 of 2ª Squadriglia at Aviano.

the Caproni as it returned and damaged it. On 15 April there was a further change of name, the unit became 2ª Squadriglia Aeroplani while the Gruppo Comando Battaglione Aviatori became IV Gruppo. On 18 April Ca 708, with pilots Verona and Canelli, bombed the station at Comen.

On 16 May 1916 two bombers, 483 and 712, took off at dawn to bomb the railroad and emplacements at Ovcia Draga. On the way back two couples of fighters attacked Ca 483 of Carta, Sala, Saletta, and Chiaperotti who fought back firing 300 rounds and claimed an enemy shot down, possibly Lloyd C.III 43.65 of Flik 2, with the crew of *Krp.* Pichler and *Fan.* Mehrfurt unharmed. In late May and June 1916, the squadron bombed targets in the northern sector of the front, where the Austrians were on the offensive, and they hit positions on Mount Erio, Val d'Astico, Roana and Campolongo. During one such actions on Val d'Astico, on 9 June a shell hit Caproni 708 wounding *cap.* Carta and gunner Saletta, but still the crew went on and completed its mission. The same plane was hit again on another mission on 16 June. On 20 June Caproni 702 and 708 bombed Roana and Rotzo in formation with ten other bombers. Austrian aircraft attacked the formation, losing one of their own.

On 18 July the 2ª Squadriglia had its first loss. Clerici, now a *capitano*, took off from Aviano to ferry Caproni 712 to Taliedo with second pilot *sold.* Francesco Sartori, Chiaperotti and mechanic Francesco Salmin. Near Poncarale, Brescia, the propeller of the central engine broke and smashed into the left fuselage, collapsing the whole airframe. There were no survivors.

On 1 August 1916 the 2ª Squadriglia at Aviano had commander Carta with pilots Sala, *cap.* Alessandro Raffaelli, *s.ten.* Lando Nugoli and Pietro Penazzi, *serg.* Canelli, Lazzarini, Primo Maggi, *cpr.* Angelo Comolli and *sold.* Ruggero Franzoni; with observers *ten.* Umberto Fioretti and Attilio Giuliani; gunners Saletta and Davide Bassotti. Its aircraft were Capronis 1234, 1243, 1244 and 1249.

In that month the unit bombed the Arsenal of Trieste, 13 September, and targets in the Carso region as part of the Sixth Battle of the Isonzo, leading to the conquest of Gorizia. In the final part of the year bad weather curtailed activities and its support to the fall offensives was limited. During the whole of 1916 the 2ª Squadriglia flew 35 war missions.

At the start of 1917 the unit was at Aviano without its commander, equipped with Capronis 1199, 1184, 1255, and 1185. It had the pilots *ten.* Leandro Negrini, *s.ten.* Sala and Lando Nugoli, *serg.* Canelli, Lazzarini, Maggi, Alfredo Alderighi and Ugo Brambilla; observers *ten.* Giuliani and Fioretti; gunners Saletta, Pietro Arlunno and Ugo Cassigoli.

The first mission of 1917 was a raid against barracks near S. Daniele del Carso on 8 January. Returining Ca 1199 suffered the attack of a fighter identified as an "Ago" that hit it repeatedly.

Above: Ca.3 of 2ª Squadriglia at Aviano showing their red circle insignia.

No war missions were then flown until March, but during a test flight on 17 March Caproni 1185 crashed, killing its pilots *serg.* Maggi and *serg.* Brambilla.

Cap. Attilio Matricardi became the new commander on 21 March 1917.

On 31 March *s.ten.* Nugoli crashed to his death on Caproni 2340 at Malpensa, where he was converting on the Ca 450 hp. On 10 April the 2ª Squadriglia was reassigned from the IV to the XI Gruppo. On 18 April the unit bombed barracks at Chiapovano, a shell from the Austrain defences went right though the wings of Ca 1199 of Negrini and Canelli.

In May 1917 the Tenth Battle of the Isonzo began: on the night of 11 May Capronis 1199, 1255, and 2611 bombed the station at San Daniele del Carso, a bomb hit a building sporting the Red Cross, where secondary explosions were seen. On 14 May, as the squadron was going to bomb positions in the Vippacco Valley, an engine seized on Caproni 1255 that crashed at take-off, injuring only the second pilot *s.ten.* Raffaele La Polla, who was flying for the first time. No injury to the rest of the crew, Negrini and gunners Cassigoli and De Lama. The 2ª received Ca.3 450 hp 2355, 2366, 2382, and 2371.

In June 1917 the 2ª Squadriglia was used on the northern front supporting the bloody attack on the Ortigara. On 10 June three bombers took off to attack Caldonazzo, but clouds hampered navigation. Ca 2382 of Matricardi and Lazzerini with Giuliani and Arlunno lost contact with the group, and bombed alone, but then faced the attack of six Austrian aircraft. Their fire wounded gunner Arlunno and put two engines out of action. Matricardi climbed in the back turret, removed the gunner and took his place, hit one of the attackers, Ca 2382 crash-landed just inside the Italian lines near Arsiero.

In early July the unit received more Caproni 450 HP. On 28 July Ca. 4040 of Matricardi, Lazzarini and Cassigoli with the observer *ten.* Gino Graziani flying to bomb Idria was damaged by anti-aircraft fire, it dropped its bombs over the Chiapovano area and returned with Cassigoli standing on the wing to patch up the radiator and one bomb dangling from its rack, which Matricardi managed to drop over an empty mountain area.

On 1 August 1917 the 2ª Squadriglia, still at Aviano with *cap.* Matricardi flew Capronis serial 2355, 2366, 2371, 4040 and 2321. Its force included pilots *ten.* Negrini, Gino Lisa and Sala, *serg.* Canelli, Lazzarini, Agostino Galli and *cpr.* Arturo Vischioni; observers cap. Attilio Giuliani, *ten.* Graziani and Giuseppe De Vizzi; gunners Arlunno, Cassigoli, Saletta, Ettore Candiani and Tarcisio Cantarutti.

In the night of 2 August 1917, the squadron attacked the arsenal of Pola. While Ca 4040 of Matricardi had to turn back for engine failure (it is to be remarked that in these occasions Italian pilots returned keeping their bombs on board) Caproni 2371 of Lisa, Vischioni and Saletta repelled the attack of an Austrian seaplane, and returned with just one engine working, crashing in a marsh, with no injury to the crew. The attack on Pola was replicated the next night. On 6 August the squadron bombed emplacements at Chiapovano with its Caproni 4040, 2366 and 2321 in close formation. A new attack on Pola took place in the night of August 9. Several actions followed in August, with even two sorties on a single day, with Capronis 4039 and 4071 often with the participation of *mag.* Armani, commander of the XI Gruppo.

On 28 August the four Capronis were tasked to bomb Vojscica. Ca 4039 had the crew of Lisa, Vischioni and Saletta, with *serg.* Galli as gunner. Three new type fighters attacked it near Cerovljie, while a single escorting Spad tried to defend the bomber. Galli was wounded, the Caproni was badly damaged and it force landed at Aiello airfield, from where it managed to return to base the same day.

In August and September, the 2ª Squadriglia was very active supporting the large 11th Battle of the Isonzo, flying 25 missions, both by day and night, with Matricardi in the lead. On 3 and 4 September, Pola was again attacked at night.

Right: A Caproni of 2ª Squadriglia at Aviano.

In September Lisa, Vischioni, Saletta and *ten. oss.* Zoppi went to Taliedo and then to Gioia del Colle to take part in the action against Cattaro, while new pilots arrived, *ten.* Colombo Conti and *cpr.* Mario Occa.

With the enemy breakthrough at Caporetto the 2ª Squadriglia was completely engaged. Two Capronis bombed Santa Lucia di Tolmino on 25 October. On 26 October all its serviceable aircraft took off to bomb the enemy at S. Lucia di Tolmino: Ca 4040 of Matricardi with *cpr.* Mario Occa, Graziani and gunner *cpr.* Egidio Porrino faced and repulsed the attack of two fighters. The bomber managed to return to base, but it was so much damaged that it was written off. Ca 4039 of Lisa, Vischioni, Saletta and gunner Carlo Tagliabue was intercepted by two fighters. Saletta was wounded, but he shot donw a figher over the Bainsizza, maybe Albatros D.III of *Vzfw.* Fritz Jacobsen, Jasta 31, but also the Caproni was doomed. The German fire wounded Vischioni twice in the chest and the plane was riddled. Lisa managed to crash land at Oleis airfield, but the Caproni was abandoned.

That was not all: also in the afternoon Ca 4057 of *ten.* Negrini, *serg.* Canelli and gunner *sold.* Candiani with *mag.* Carta as observer attacked the same target, and once again two fighters engaged it. One of them was hit and glided away, while the other one quit the field, but the Caproni was badly damaged and written off. In the night 4071 with Armani and Ridolfi bombed Grahovo

The squadron attacked bridges, railroad stations and the roads used by the advancing Austrians and Germans, and dropped food for the Italian troops, then at the end of October it moved back, first to Arcade and then, on 9 November, to Ghedi. During the ferry flight to Arcade Ca 4174 suffered a double engine failure and crashed near Vigonovo, with no injury for its crew. The Capronis serials 2321, 2355, 4040, 4057, not flyable, were burned at Aviano, 4174 and 4039 were abandoned at other airfields, 2401, 4071, 4144, 4180 and 4191 were flown to safety.

On 14 November the 2ª Squadriglia was assigned to the XIV Gruppo, of commander *mag.* Vittorio De Muro. At that time, it had Capronis 4170, 4180 and 4191. Its force consisted of commander Matricardi, pilots *ten.* Colombo Conti and Raffaele Coppola, *s.ten.* Lisa, *m.llo* Baldassarre Cucchiara, *serg.m.* Guido Colli and *cpr.* Mario Occa; observers *ten.* Graziani and Pietro Bassi; gunners Candiani, Tagliabue, Luigi Assier, Igino Ferraretti and Egidio Porrino.

On 15 November three Albatros fighters of Flik 55/J intercepted Caproni bombers from different squadrons as they returned from an attack against railroads at Levico and Caldonazzo. Julius Arigi, Josef Kiss and Josef von Mayer, profiting from the lack of escort fighters, attacked Ca 4191 of Lisa and Colli with *ten. oss.* Bassi and gunner *cpr.* Tagliabue. The bomber was shot down with the death of its crew. Lisa was posthumously awarded the Medaglia d'Oro al Valor Militare, Gold Medal, while the other airmen got a Silver Medal. Arigi before shooting down 4191 had attacked 4169 of the 10ª Squadriglia which crash-landed with gunner Natale Coccè killed. On 22 November there was the last combat flight of 2ª Squadriglia in 1917.

On 5 December the Command of the XIV Gruppo

Left: The Capronis that were burned during the enemy offensive I October 1917.

moved from Ghedi to Padua, the squadron remained at Ghedi, in a Ghedi Airfield Detachment under *cap.* Barucchi until the end of the year, then it also moved to Padua. During 1917 the unit had flown 139 combat missions.

In January 1918 the unit, based at Padua under *cap.* Matricardi, had pilots *m.llo* Cucchiara, *serg.m.* Riccardo Salino, *serg.* Occa, observer *ten.* Gino Graziani, gunners *serg.* Egidio Porrino and Ettore Candiani. Its squadron markings were red circles, from one to four, on the fuselages.

It was active in the sector of the Asiago plateau. On 5 January Matricardi on Ca 4226 with personnel from the 7ª Squadriglia bombed the airfield of La Comina. On 12 January Ca 4226 of *cap.* Matricardi, *ten.* Coppola, *cap. oss.* Graziani and *serg.* Porrino raiding the railroad at Primolano straggled from the group and three fighters attacked it. Porrino was killed, the bomber was damaged but Graziani kept the fighters at bay, until *serg.* Andrea Teobaldi of the 82ª Squadriglia, flying Hanriot 6127 "A" drove the enemies away. The bomber crash-landed near Montebelluna, with no further injury for its occupants.

On 13 January the command of the unit passed from Matricardi to *cap. pil.* Leone Cola De Renzis.

In the night of 2 February Ca 4193 of De Renzis and Cucchiara bombed Mount Rovere.

On 5 February King Victor Emmanuel visiting Padua passed in review the squadron. Activities were reduced in February and March. On 20 February, De Renzis and Occa on 4193, the only serviceable Caproni, bombed La Comina, three days later the same crew bombed the station of Bolzano. On 5 March at noon De Renzis and Occa bombed barracks at Oderzo, on the 21st they failed to attack La Comina on account of fog, and returned carrying their bombs, they tested a new wireless device, that was correctly picked up by the ground station at Padua.

Three bombers attacked Primolano by night on 28 March, activities stopped altogether in April, as weather was bad and only one airplane was flyable. On 7 April *cap. pil.* Vittorio Bergonzi became the new commander. New pilots joined the squadron: *m.llo* Empedocle Piccinini, *serg.m.* Rinaldo Salino and *cpr.* Francesco Arbore and new aircraft like Ca 11491.

Operations resumed on 4 May with an action near Alano. Ca 4193 couldn't reach the altitude and dropped its bombs on Garda Lake.

On 22 May Ca 11491 with pilots *ten.* Domenico Giuriati and *serg.* Vischioni, back in service after having been badly wounded on 26 Oct. 1917, observer *ten.* Francesco Pistocchi and gunner Francesco Zavattaro attacked the airfielfd at Motta di Livenza but it suffered an engine failure on landing and crashed near its base, with its crew uninjured. In June a few leaflet dropping missions were flown.

During the battle of the Piave the squadron was in action, flying the old Ca 2314 and 2404 from the disbanded 16ª Squadriglia, that also transferred its pilots *ten.* Eugenio Pasquino and Tommaso Salazar and *serg.* Luigi Pò and observer *ten.* Sigismondo Hirsch. The first American pilots trained at Foggia arrived: Lieut. Harold Holtz and Clarence M. Young. Young, holder of United States Civil Air License No. 2, later became Assistant Secretary of Commerce for Aeronautics and Pan American World Airways executive who directed establishment of the first flying routes across the Pacific.

On 20 June a Caproni (serial unknown) failed to return. The fate of its crew, Lieut. Young and *ten.* Eugenio Pasquino, observer *ten.* Silvio Tedesco and gunner Secondo Fassio, was told in a note by Young, that an Austrian airplane dropped: the pilot said that he was disoriented in the clouds, landed near Cà Pescatori and was captured.

On June 24 attacking the station at Pramaggiore the Caproni of *ten. oss.* Gaetano Battiati and Salazar with gunner Carlo Surniotti was badly mauled by flak, it made an emergency landing at Padua airfield breaking its undercarriage. Salazar was slightly wounded.

On 25 June gunner Giuseppe Testa, on board Caproni 4176 with pilots Cucchiara and *serg.m.* Rinaldo Salino, and *ten.* Raffaele Di Caudia claimed a fighter shot down after a mission against a munitions dump near Orsago. The 2ª Squadriglia had an odd mix of Capronis, the old 2401, the brand new 11498 and 11506, the well-worn 4176.

On 17 July this unit took part in the big combined

Above: Spectacular crash of a Caproni; unfortunately, the details are not known.

operation against Pola with Ca 4180 and 11506 of Bergonzi and Holtz, while 2401 had to turn back due to its faulty engines. Ca 11506 was back in action the same day, now with Cucchiara and Salino, dropping leaflets across the Piave River. Ca 2401 went into action against the railroad at Pordenone in the night of 20 July, with pilots Salazar and Angelo Pizzini, from 10ª Squadriglia, and *magg.* Vece, commander of XIV Gruppo, the mission was successful but

Right: Clarence M. Young, (right), Director of Aeronautics, Department of Commerce, in 1927 delivering his pilot's licence to Santa Claus.

the bomber was damaged by Italian artillery over Treviso.

On 22 July the 2ª Squadriglia moved to Arquà Petrarca. At the end of July the unit had Caproni 2401, 2404, 4176, 4180, 11495 and 11506. Its force included pilots Lieuts. Harold F. Holtz, Frank Lambert, Patton Mc Gilvary, *ten.* Vittorio Mammoli, Tommaso Salazar, *s.ten.* Francesco Vecchio Verderame, *m.llo* Cucchiara, *serg.m.* Salino and Lorenzo Lena, *serg.* Luigi Pò and *soldato* Gildo De Santis; observers *ten.* Pistocchi, Di Caudia, Elio Turchetti and *s.ten.* Edoardo Cipparone; gunners Candiani, Italo Soprani, Carlo Agnello and Antonino Crifò.

In August the unit attacked enemy airfields at Feltre, Portobuffolè, Comina, Casarsa and Codega di S. Urbano, and also dropped leaflets. On 1 Sept. Ca 4180 with the American pilots, Lts. Leland P. Spore and McGilvary, taking part in a mass attack on Aviano had its radiators wrecked and two engines out of action, but the pilots managed to land safely at Padua.

On 10 September *cap.* Bergonzi returned as commander, and remained until the end of the war. A new American pilot arrived, Lt. Jay W. Blackman.

On 7 October Caproni 11506 loaded with bombs was destroyed by fire at Arquà Petrarca. The fireman *sold.* Leone Sabbatini tried to stop the fire from reaching the other planes and was killed when the bombs exploded, a Silver Medal was the prize for his gallantry.

When the final offensive was launched, the 2ª Squadriglia was active against enemy airfields and lines. On 27 October the crew of Vecchio Verderame, Lena, Pistocchi and Agnello on Ca 4180 bombed a munitions dump at Sacile. Five fighters attacked just after the bombs were dropped, and in a long combat Agnello, wounded in a hand, claimed a fighter shot down. The Caproni was damaged, Pistocchi was wounded and the Caproni managed to land at the British airfield at Treviso. Caproni 2401 and 11495 flew the last mission of the war, bombing troops along the Sacile-Pordenone road both in the morning and in the afternoon of 30 October.

In 1918 the 2ª Squadriglia flew 88 war sorties, and its total for the conflict was 289. At the armistice the unit was at Arquà Petrarca, with commander *cap.* Bergonzi, flying Capronis 2401, 4180, 11495 and 11506. Its pilots were Lieuts. Mc Gilvary, Jay V. Blackman and Leland P. Spore, *ten.* Mammoli, *s.ten.* Vecchio Verderame, *m.llo* Cucchiara, *serg.m.* Lena, Po, Salino, serg. Umberto Merlino; observers *tenenti* De Candia, Turchetti, Gaetano Battiati, Vittorio Centa and Giovanni Vergani; gunners: Agnello, Soprani, Testa and Silvio Santini.

In February 1919 the 2ª Squadriglia was assigned to the IV Gruppo, but one month later it was disbanded. Its aircraft were Ca.3 2404, 4058, 4180, 11495 and 11497, and Ca.5 11545.

Casualties and Combat Losses of the 2ª Squadriglia

Date	Place	Aircraft	Crew	Cause
18.7.1916	Brescia	Ca 712	*Cap.* Clerici KIC, *sold.* Sartori KIC, *sold.* Chiaperotti KIC, Salmin KIC	Accident
10.3.1917	Aviano	Ca 1185	*Serg.* Brambilla KIC, *serg.* Maggi KIC	Accident
31.3.1917	Malpensa	Ca 2340	*S.ten.* Nugoli KIC	Accident
10.6.1917	Caldonazzo	Ca 2382	*Cap.* Matricardi, *serg.* Lazzerini, *ten.* Giuliani WIA, *serg.* Arlunno WIA	Fighters
28.8.1917	Cerovljie	Ca 4039	*S.ten.* Lisa, *cpr.* Vischioni, *cpr.* Saletta, *serg.* Galli WIA	Fighters
26.10.1917	S. Lucia	Ca 4039	*S.ten.* Lisa, *serg.* Vischioni WIA, *cpr.* Tagliabue, *cpr.* Saletta WIA	Fighters
26.10.1917	S. Lucia	Ca 4040	*Cap.* Matricardi, *cpr.* Occa, *ten.* Graziani, *serg.* Porrino	Fighters
26.10.1917	S. Lucia	Ca 4057	*Ten.* Negrini, *serg.* Canelli, *sold.* Candiani, *magg.* Carta	Fighters
15.11.1917	Arsiero	Ca 4191	*S.ten.* Lisa KIA, *serg.m.* Colli KIA, *ten.* Bassi KIA, *cpr.* Tagliabue KIA	Fighters
12.1.1918	Cismon	Ca 4226	*Cap.* Matricardi, *ten.* Coppola, *ten.* Graziani, *serg.* Porrino KIA	Fighters
20.6.1918	Piave	Caproni	Lt. Young POW, *ten.* Pasquino POW, *ten.* Tedesco POW, *sold.* Fassio POW	Accident
27.10.1918	Sacile	Ca 4180	*S.ten.* Verderame, *serg.m.* Lena, *ten.* Pistocchi WIA, *sold.* Agnello WIA	Fighters

3ª Squadriglia

Above: Caproni 300 hp 1129 of 3ª Squadriglia starting for a mission at Aviano.

The 3ª Squadriglia was formed at Malpensa on 8 October 1915 with pilots *cap.* Luigi Carnevale, *cap.* Giulio Palma di Cesnola and *serg.m.* Luigi Bertello. Its commander was *cap.* Pier Ruggero Piccio and its equipment consisted of Caproni 300 hp serials 482 and 486. Observers were *cap.* Pico Deodato Cavalieri and *brig.* Giuseppe Marini. The unit moved to La Comina on 17 October, assigned to the Gruppo Comando Battaglione Aviatori.

On 19 October it received Caproni 702 with its crew of pilots *cap.* Ernesto Pellegrino and *s.ten.* Ettore Marazzi and the same day the three bombers attacked Aisovizza twice. They flew seven more missions in October, each time with three planes. On 5 November Pellegrino and Marazzi were assigned to the flying school of Cascina Malpensa, while *ten.* Giampiero Clerici and *s.ten.* Alberto Verona replaced them. Until the end of 1915 the unit operated in the sector of the Carso, flying 26 bombing missions in October, 15 in November and 3 reconnaissance flights in December. Ca702 went back to the factory and 705 replaced it. A new pilot was *cpr.* Lodovico Montegani, later test pilot with Breda. Only one bombing mission was flown in January 1916, on the 17th Piccio and Palma flew on reconnaissance to Fiume.

On 18 February 1916 the 3ª Squadriglia sent its Ca 482 in the raid against Lubjana. This bomber, with pilots *cap.* Palma and Carnevale, dropped twenty 162-mm bombs on the railroad station, but on the way back it suffered the attack of a two-seater identified as an "Aviatik". The return fire from the bomber possibly wounded the Austrian gunner, and the Caproni returned with many bullet holes.

On 1 March *cap. pil.* Adolfo Resio, coming from the 6ª, replaced Piccio as commander. While bad weather curtailed operations, the wing struts were modified and machine guns were fitted to the bombers. Armor for the pilots was also tested. When flying was possible, the squadron tested flight formations for better mutual protection.

On 2 April Resio, Palma, Marini and gunner Ugo De Ciantis flew Ca 486 on a reconnaissance to Adelsberg, but they had to return because exaust flames burned a wing strut, that then collapsed. Both Ca 482 and Ca 486 were withdrawn from service. On 15 April 1916 the unit received the new name of 3ª Squadriglia Aeroplani while the Gruppo Comando Battaglione Aviatori became IV Gruppo Aeroplani. On that day the unit received Caproni 1159, which was damaged in a landing and struck off charge three days later. On 26 April Ca 705 of Carnevale and Marini with *serg. pil.* Angelo Mamoli and gunner Azeglio Dini raided the station at Ovcia Draga. An Austrian shell damaged the controls and the left engine, and the bomber force landed at Bertiolo, from where it was later recovered.

In May and June the 3ª Squadriglia operated in the sector

Above: Pico Deodato Cavalieri and Cola Leone De Renzis. Cavalieri, a Jew, died in training as a seaplane pilot. His family gifted their palace in Ferrara to the nation; called *"Casa della Patria"*, it is the home to veterans' associations.

of the Plateaus against the Austrian offensive, dropping steel darts. On 16 June an Austrian airplane attacked Ca 1129 near Mount Longara, but it was driven away, possibly hit by defensive fire.

In June *cap.* Resio was called to the Technical Office, and *cap.* Carnevale became acting commander, until July, when he was sent to 6ª Sq. and *cap.* Luigi Falchi took over as commander and Palma was re-assigned. At this time 3ª Squadriglia consisted of pilots *cap.* Palma, *s.ten.* Bertello, *serg.* Felice Strinasacchi, Angelo Mamoli, *cpr.* Quintilio De Medio, gunners De Ciantis and Antonio Turola, with Capronis 1129, 1146, 1149, 1156.Three bombing mission were flown in April, one in May, 9 in June, all in the northern sector, two in July.

On 1 August 1916 the squadron attacked the torpedo factory at Fiume, sending three bombers (1156, 1129 and 1246) out of a total formation of 25. On 9 August gunner De Ciantis in Ca 1129 claimed another Austrian airplane during a raid against the railorad stations of Dornberg and Prvacina. In this period the squadron operated supporting the sixth offensive of the Isonzo, that lead to the conquest of Gorizia and saw the use of massive formations of Capronis. On 15 August the 3ª Squadriglia, now based at Aviano flew the Capronis 1129, 1156, 1254 and 1246: its pilots were *cap.*

Above: Ca.1 or 300 hp 1136 of 3ª Squadriglia with a white cross emblem, probably the Cross of Savoy, I a red field with a blue border.

Palma di Cesnola, *ten.* Germano Ruggerone, *serg.m.* Bertello, *serg.* Mamoli and Strinasacchi, *vicebrigadiere* Demetrio Artuso, *cpr.* De Medio and *cpr.* Mario Pereno; observers: *cap.* Cavalieri, *ten.* Tito Giovambattista Benetti and *ten.* Giovanni Venditti; gunners De Ciantis, Angelo Balzarotti and Antonio Turola. Ruggerone, an aviation pioneer and a veteran of the Libyan war, used to be a famous bycicle racer with the name "Eros". On 26 August *cap.* Luigi Falchi became the squadron commander.

On 13 September all four aircraft of the 3ª Squadriglia bombed the Arsenal of Trieste, facing the enemy seaplanes; anti-aircrat shells wounded gunner De Ciantis. On 6 December its four 350 hp bombers attacked the roads around Schoenpass and on 10 December it bombed the railroad at San Daniele del Carso, despite the terrible weather.

On 1 January 1917 the 3ª Squadriglia, at Aviano, with commander Falchi, had Capronis serials 1185, 1196, 1246 and 1254. Its pilots were *ten.* Ruggerone, Casimiro Buttini, *s.ten.* Luigi Sambo, *v.brig.* Artuso, *serg.* Mamoli, Strinasacchi and Giovanni Albano; observers *ten.* Benetti and Venditti; gunners: De Ciantis, Alberto Farneti, Antonio Garrone and Enrico Guarinoni.

Right: After the tragic results of the raid of 18 Februay 1916, armor was tested for the Caproni airmen, but it was not adopted.

On 6 January Falchi, Ruggerone and Artuso flew Ca 1195 to bomb Trieste by night, but due to the weather and problems with the bomb shackles they dropped the bombs over the Hermada Hill. It was the first night sortie, carried out without permission from higher commands, for which Falchi was reprimanded. On 8 January two bombers attacked S. Daniele railroad station. Five fighters attacked Ca 1254 of Ruggerone, Arturo, and Benetti and De Ciantis, damaging it and putting one engine out of action; an escorting Nieuport fighter arrived, after many attacks one of the Austrians was apparently driven down and the others retreated. Weather stopped further operations; meanwhile *cap.* Falchi went on leave, being promoted and assigned as commander of the IV Gruppo, and he did not return to the squadron. On 10 April 1917 the 3ª Squadriglia was assigned to the new XI Gruppo. On 25 April Benetti, acting commander, was assigned to the 7ª Squadriglia while Ruggerone became commander of the 3ª, which also received *cap.* Cesare Darby who became its commander on 25 April.

On 3 May 1917 two bombers attacked Sesana despite a dense haze. Ca 1196 failed to find its target, dropped the bombs blindly and returned to base while Ca 1246 of Sambo, Albano and De Ciantis got lost and force landed in the Po River near Massa Superiore, its crew swimming to safety.

The squadron attacked Austrian positions during the 10th Battle of the Isonzo and bombed Birhula, Vogersko and Vojscica. On 23 May, on its way to Birhula, Ca 1196 was attacked first by an "Ago" and then by a "Fokker". The latter was probably hit and driven away; the Caproni was badly damaged but returned to La Comina. In June this unit operated in the area of the Asiago plateau, supporting the Battle of the Ortigara.

Its force at this time consisted of pilots *cap.* Ruggerone and *ten.* Casimiro Buttini, *s.ten.* Vito Pugliese, Livio Sambo and Bernardo Viano, *v.brig.* Artuso, *serg.* Mamoli, Albano and Silvio Geri, observers *ten.* Venditti, Weiss Poccetti and Pascotti, gunners *cpr.m.* Guarinoni and *soldati* De Ciantis, Ermenegildo Savo, Cesare Vernizzi, Antonio Garrone, Farneti, Raffaele Negroni, Bruno Taiuti, Fadigato. Ca 450 hp replaced the Ca 300 hp.

On 22 July a group of high officers from the United States, with French, Japanese, Spanish and Serbian observers visited the airfield. On the occasion there was a static display with aircraft from other airfields and two airships from Casarsa aerodrome, one of which carried Gen. Maggiorotti, commander of the Italian air service. Medals were awarded and Gabriele d'Annunzio spoke.

On 28 July the bombers attacked the mercury processing plant at Idria, while contested by enemy fighters. On 1 August 1917 the 3ª Squadriglia of the XI Gruppo at Aviano had Capronis 2357, 2367, 2388 and 2394. Under its commander *cap.* Ruggerone there were pilots *ten.* Buttini, Bernardo Viano and Vito Pugliese, *s.ten.* Sambo, *serg.* Albano, Mamoli, *v.brig.* Artuso and *sold.* Riccardo Casellato, observers *ten.* Venditti and Weiss Poccetti; gunners De Ciantis, Farnetti, Garrone, Raffaele Negroni and Cesare Vernizzi.

In the night of 3 August four bombers, tasked with the bombing of Pola naval base, took off, but only one reached its target, while the other ones had to abort the mission. Ca 2367 of Viano and Albano, with Venditti and Negroni, dropped eight bombs on the Arsenal causing a fire. On the way back, having to make detours to avoid clouds, the bomber ran out of fuel and force landed near Spilimbergo, but the crew got out of the wreck unharmed. The next night the three remaining bombers returned to the same target, two of them attacked successfully while Ca 2357 of *serg.* Casellato first suffered the attack of an enemy seaplane, possibly A.11 piloted by Gottftried Banfield, and then had to return at low level on account of an engine failure.

From mid-August to September the 3ª Squadriglia, often enjoying fighter escort, supported the huge offensive of the 11th Battle of the Isonzo. On 9 September, during an action against Ternova in the final phase of the battle this squadron suffered its only loss due to enemy action. Anti-aircraft fire hit Ca 4060 of *ten.* Buttini with *serg.* Luigi Remitti, Poccetti and Farnetti and beheaded second pilot Remitti. Buttini nursed the airplane to the Italian lines and to a forced landing on the slope of mount Peuma. During the night the body of the dead airman was recovered. King Victor Emmanuel III on his own initiative assigned a *Medaglia d'Oro al Valor Militare* to Buttini and a Silver medal to the other crew members.

After the events of Caporetto the 3ª Squadriglia burned at Campoformido its unserviceable Ca 4068 and then on 31 October it moved to Padua with Capronis 4037, 4079, 4141 and 4172. At this time with *cap.* Ruggerone there were the pilots *cap.* Carnevale, *ten.* Buttini, and Vittorio Bergonzi, *s.ten.* Sambo, *aspirante* Artuso, *serg.* Giacomo Armano and Pietro Zemiti, *sold.* Casellato; observers *ten.* Ezio Sbrozzi, Poccetti and Venditti; gunners De Ciantis, Farneti, Negroni, Vernizzi and Giulio Acrocca.

On 2 November the unit moved again, to Arcade. On landing Ca 4141 had an accident and was wrecked, but its crew was unharmed. On 6 November the 3ª Squadriglia was ordered to move again to Ghedi, where it was to be re-equipped. On 13 November the unit was assigned to the new XIV Gruppo Aeroplani.

During that troubled time the unit flew many missions to slow down the enemy offensive. On 5 December it was moved once again, returning to Padua. 16 successful bombing sorties were flown in December. On 20 December it was tasked with a mission at S. Donà di Piave. This action deserved a citation from the Command of the Bomber *Raggruppamento*, but it was the last one, as the 3ª Squadriglia was ordered to move to the French front as part of the new XVIII Gruppo, together with the 14ª and 15ª. On the 24th it turned part of its material to the 10ª Squadriglia and shipped the rest to Taliedo, where it would get new airplanes. The personnel followed two days later by railroad. In 1917 the 3ª Squadriglia had flown 142 bombing sorties.

At the start of 1918, its personnel consisted of *cap.* Ruggerone, *cap.* Vito Pugliese, *ten.* Livio Sambo, *asp.*

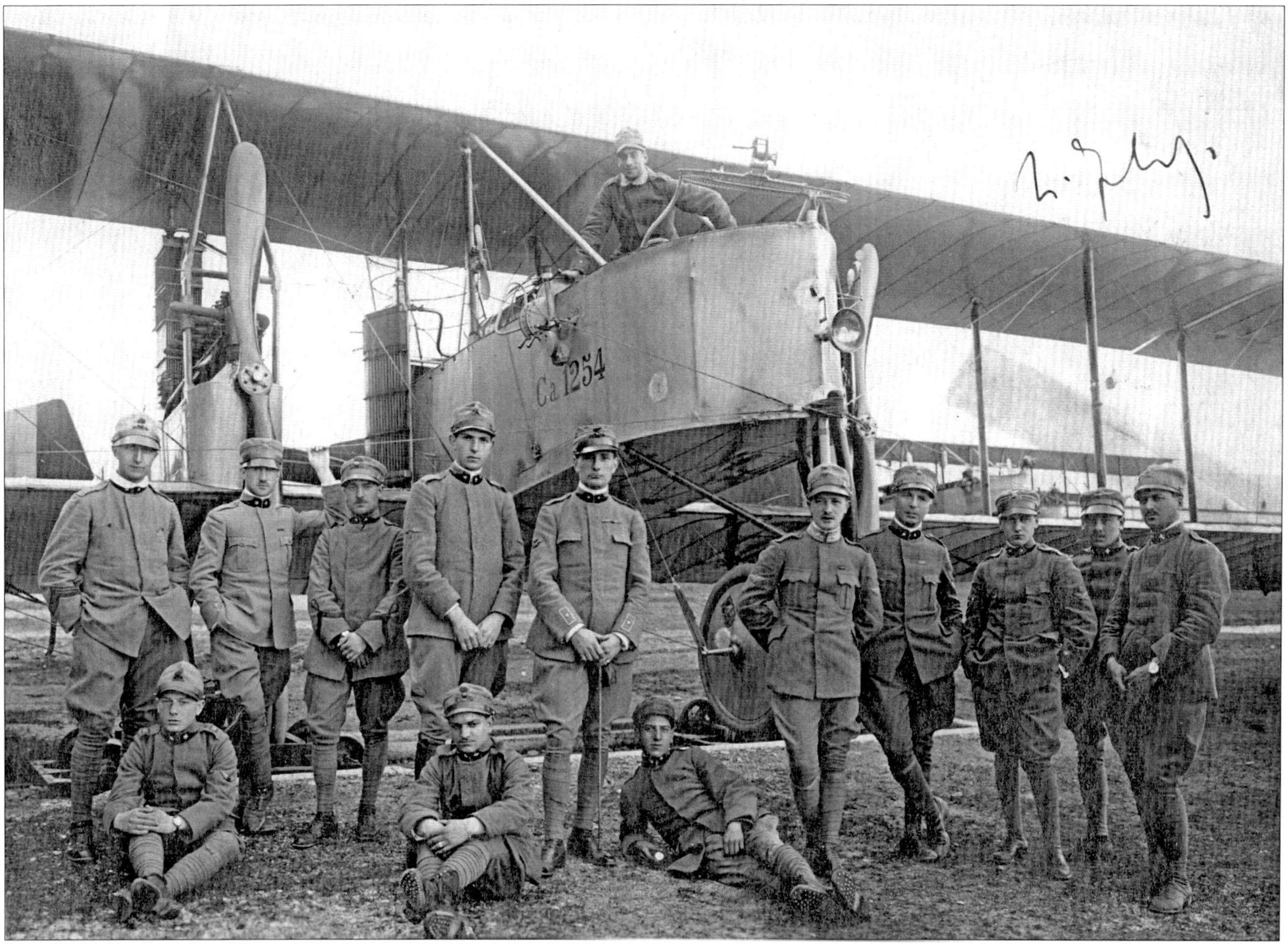

Above: Personnel of 3ª Squadriglia with Caproni 350 hp 1254. The second man from the left standing is *ten.* Casimiro Buttini, the fifth is *magg.* Falchi.

Demetrio Artuso, *sergenti* Giovanni Albano and Pietro Zemiti, *caporali* Giacomo Armano, Riccardo Casellato, Eugenio De Marchi. The observers were *tenenti* G. Battista Pratesi, Luigi Mancini, Ezio Sbrozzi, Giovanni Venditti, with gunners *caporali* Osvaldo Clerici, Ugo De Ciantis, Raffaele Negroni, *soldati* Giulio Acrocca and Cesare Vernizzi.

On 9 January 1918 the 3ª Squadriglia, fully equipped with men and material, went by rail to France. Two days later it began to prepare its equipment at the airfield of Dijon. Ruggerone brought to France Caproni serials 2363, 2378, 4219 and 4229. Bad weather precluded any flying until 9 February, when two airplanes of the group, one from the 3ª and one from the 15ª, dropped flowers over the city hall of Dijon in the course of a decoration ceremony. On 19 February the 3ª Squadriglia moved to Ochey airfield, Department of Meurthe et Moselle, the whole XVIII Gruppo was assigned to *Escadre* 11, reporting to the Aviation Command of the Group of Armies East. Bad weather again precluded any flying until 17 March when *cap.* Renato De Riso, commander of XVIII Gruppo, and Artuso on Ca 2363 and Ruggerone and Albano on Ca 4229 bombed the German airfield at Metz Frescaty. The action was repeated on 23 March by three bombers. On 24 March four Capronis bombed the station at Thionville.

On 28 March the French command ordered a movement to Villeneuve airfield, to support the offensive of the Champagne. Due to the weather, the Italian bombers could move only on 3 April.

On 12/13 April was the next mission, a night sortie against Montcornet station for two Caproni, 4219, pilots Ruggerone and Albano, and 4229. Returning, Ca 4219 met a thick fog and crashed on landing at Longvic. Second pilot Albano was unhurt, Ruggerone and gunner De Ciantis were injured, while observer Venditti was killed.

In the spring and summer 1918 the 3ª Squadriglia successfully attacked targets at Leons, 3 May, 2363 and 2378, and different attacks against raildroads, by one or two planes, in the nights of 14, 15, 20. 22, 27, 28, 29 and 31 May.

On 1 June 1918 commander *cap.* Ruggerone, pilots *ten.* Artuso, Sambo, Viano, *serg.* Albano, Armano, Zemiti, *cpr.* Eugenio De Marchi, *sold.* Casellato; observers *tenenti* Pratesi, Sbrozzi, Rinaldo Parodi, Pietro Scotti Fantoni and Mario Verdoia; gunners Acrocca, Clerici, De Ciantis, Farneti, Negroni and Vernizzi formed this unit.

On 5 June Ca 4236 of Ragnelli and Pica attacked the station at Soisson by day. In June the 3ª Squadriglia flew 16

Above: Lined up for a ceremony at Aviano, July 1917, the Capronis of 3ª Squadriglia and behind them those of 1ª Squadriglia.
Below: Ca.3 2367 crashed in the return from the bombing of Pola, 5 August 1917.

Left: The Ca.3 4060 of *ten*. Buttini shot down by ground fire on 9 September 1917.

Above: The poet Gabriele d'Annunzio with other officers with a Caproni of 3ª Squadriglia.

Above: Ca.3 4229, protagonist of many night raids over the German lines, in France.

night bombing missions against railroads by a single Caproni, and a mission of two, Ca 4048 and 4236, in the night 29/30 June.

In July there were 31 sorties, with two or three bombers every night. In the night of 20 July *s.ten.* Artuso flew Ca 4229 twice, once with *ten.* Sambo and once with *serg.* Zemiti, attacking two different targets.

By this time the Germans had identified the airfield, and from then on, they bombed it every day, so that on 5 August the Italian Group moved to Chernizey airfield in the Department of Vosges. The French summer brought rain and bad weather, and many actions were aborted. Eight sorties were flown in August, nine in September, mainly against railroad targets. In the night of 16 September Ca 4236 with the same crew, Viano, Armano, Pratesi and Negroni flew two missions.

In the night of 10 October, the Capronis dropped 1,386 kgs of bombs over Audun le Roman, in a total of 14 missions by one or two bombers, four Capronis flew in the night of 30/31 October bombing Longuyon.

By the end of October, the 3ª Squadriglia, now equipped with Capronis 2366, 4048, 4229 and 4236, received personnel from the 15ª that was repatriated: *tenenti* Vincenzo Biggio, Alessandro Borello, Federico Molteni, the top ace of bomber units, Stefano Righi and *serg.* Luigi Vallerga. Besides these pilots, the unit also got observers *ten.* Giorgio Morigi and *s.ten.* Dolores Lora Tonino with gunners Vincenzo Cannoni and Luigi Carletti.

The final action of the war was a mission against Longuyon in the night 31 October/1 November with four Capronis in action. On 10 November, as the unit was ready to attack Audun le Roman, it received news of the Armistice between the Allies and Germany.

In early 1919 the unit was at Villacoublay engaged in experiments of air mail. The 3ª Squadriglia never returned to Italy, as it was disbanded in France on 6 November 1919.

Casualties and Combat Losses of the 3ª Squadriglia

Date	Place	Aircraft	Crew	Cause
9.9.1917	Ternova	Ca 4060	*Ten.* Buttini, *serg.* Remitti KIA, *ten.* Poccetti, *sold.* Farnetti	Ground fire
13.4.1918	Longvic	Ca 4219	*Cap.* Ruggerone WIC, *serg.* Albano, *ten.* Venditti KIC, *serg.* De Ciantis WIC	Accident

Above: Ca3 4229 carried a peculiar insignia, the rank of a sergeant, as if meaning that also a NCO could command a bomber, with an eagle carrying a bomb.

4ª Squadriglia

Above: The insignia of the 4ª Squadriglia Caproni.

Above: *Cap*. Filippo Ungania, the gallant commander of 4ª Squadriglia.

This unit was formed on 17 October 1915 at Pordenone as 4ª Squadriglia Aeroplani Caproni, assigned to the Supreme Command (Comando Supremo dell'Esercito Mobilitato) and reporting to the commander of the Battaglione Squadriglie Aviatori, *magg.* Alfredo Barbieri. Its force included commander *cap. pil.* Dario Ungania, pilot *serg.m.* Enrico Mazzetti, gunner Emilio Saletta, observers *cap.* Giacomo Agosti and *ten.* Rivolta. Originally it had Caproni 300 serial 488, and before the end of the year it received 703 and 704.

It flew its first mission on 19 October sending a bomber against Aisovizza airfield, repeating this action on 20, 21 and 22 October. In late 1915 this unit flew against targets in the Carso region, during the 3rd and 4th Battles of the Isonzo. On 5 November *cap. pil.* Tullio Visconti and Gaetano Turilli joined the unit. On 14 December a Caproni made a reconnaissance to Trieste, Fiume and Buccari with *cap.* Ungania and *cpr.* Gaudenzio Signorelli and *TV* Paolo Vandone, a naval officer observer. On 15 December the unit moved to Aviano and got its new designation as 4ª Squadriglia da offesa (Ca). It flew 13 combat sorties in 1915, 11 of them with Ungania, the last one a reconnaissance on 21 December with *ten.* Ferruccio Capuzzo.

In January 1916 the 4ª Squadriglia, based at Aviano, had the same force. Three actions were flown on 9, 13 and 14 January, the last one with two airplanes. On 18 February there was the first loss. During the bombing raid against Lubjana, flown by Ca 488 and 703, two Fokker E.III's with pilots *Hpt.* Kostrba and *Oblt.* Hautzmayer attacked Ca 703 of Visconti and Turilli. Other Austrian aircraft joined the fight and the Caproni was forced down near Merna. Visconti was dead, *cap.* Turilli became a prisoner of war. Kostrba had already engaged the Caproni of *cap.* Salomone and a Caudron.

On 15 April, the unit got the new designation of 4ª Squadriglia Aeroplani. At that date its force consisted of pilots *cap.* Ungania, commander, Mazzetti (now a *s.ten.*), *s.ten.* Valentino Zanini, *serg.* Mario Borghi and gunners Arlunno, Manieri and Visioli, flying the two Ca 300 serial 488 and 704 and a new Ca 300 serial 1134. In May the unit received pilot *cpr.* Francesco Caparello and gunner Eugenio Visioli.

The squadron returned to action only on 2 April 1916, when Ungania and Mazzetti attacked Adelsberg, on the 11th, with Mazzetti a patrol flight took place. On 20 April two crews, Ungania and Mazzetti, Vanini and Borghi, attacked the seaplane station at Trieste. The latter crew flew a protective patrol over Udine on May 5, Ungania and Mazzetti bombed Tolmino on 10 May. Two bombers

Above: The Austrians repaired Caproni 703, shot down on 18 February 1916, and flew it at Aspern. Here it is seen with Brandenburgs 26.41 and 26.45, still carrying the Italian colors on its rudders. (Flygvapnetmuseum)

Above: Ca3 1197 turned over on landing on 14 May 1917 after bombing Chiapovano. At that time the unit used for identification a system of white, red, and blue squares in different combinations. (Archive Caliaro)

Above: Caproni 300 hp 1183 according to documents was assigned to 6ª Squadriglia, but here it carries the blue, red, and white squares, markings of 4ª Squadriglia. (Archive Quelli del 72)

attacked Kostanjevica on May 16. Then actions shifted to the North on account of the Austrian offensive, Mount Erio was the target on 30 May, with Ungania now flying with Caparello, then nine missions were flown in June 1916, including an attack on Cirè airfield on June 20 during the Italian counteroffensive in the Asiago plateau.

On 25 June 1916, Ca 704 and 1134 bombed Albergo Ghertele, Austrian aircraft attacked Ca 704 and damaged it badly. As the bomber tried to land at Bassano airfield it crashed, killing pilots *cap.* Ungania and *s.ten.* Mazzetti and gunner Visioli. Only a Silver Medal was assigned to Ungania, a very active and daring squadron commander.

Command of the 4ª Squadriglia was then assigned to *cap. pil.* Luigi Carnevale. On 1 August 1916 the unit had pilots Carnevale, *s.ten.* Luigi Signorini and Zanini, *m.llo* Giuseppe Farina, *serg.* Borghi, *cpr.m.* Loreto Mazzola, *caporali* Pietro Asteggiano, Renzo Barrella and Francesco Caparello, *sold.* Edoardo Avenati; observers *ten.* Giulio Porta and *s.ten.* Ugo Degani; gunners *caporali* Manieri, Sbaraglia and *sold.* Angelo Ferraris and Schirio.

There were three Capronis, 1228, 1248 and 1134 on strength. On 1 August they all bombed the Whitehead torpedo works at Fiume. *S.ten.* Signorini, *cpr.* Caparello and gunner Sbaraglia were flying Ca 1228 and the Austrian ace Gottfried Banfield, flying Lohner M L16 attacked them. Caparello was killed, hit in the head, and the bomber crash landed in a fir grove. The survivors were captured; Signorini was repatriated in 1917 in a prisoners exchange, on account of his illness.

Just five days later, on 6 August 1916 the Squadriglia had more casualties. Austrian aircraft over Miramare, Trieste, attacked Caproni 1134, crewed by Zanini, Borghi and Manieri as they returned from an attack against the Opicina rail station. Defensive fire from the bomber hit the engine of Lohner L 99 and forced it to drop out of the fight, but the other seaplane, L 16 of Banfield, pressed on its attack, seriously wounding Zanini and putting two engines out of action. The Caproni force landed between Sistiana and Nabresina, *s.ten.* Zanini died, while Manieri and Borghi, both injured, were captured. The chivalrous Banfield dropped a message over the lines, reporting the lot of the Italian crews, with photographs showing the military funeral of Zanini.

Activity of this Squadriglia was intense all through the summer, as it attacked enemy communication lines and airfields. On 19 August Ca 1248 turned over in a landing after a test flight, and was completely destroyed, but its crew of *s.ten.* Ezio Massa and *cpr.m.* Mazzola, observer *s.ten.* Degani and gunner Ferrari, survived.

In the late summer 1916 new pilots and aircraft arrived to make up for losses, thus on 1 October the unit had as pilots Carnevale, commander, *cap.* Giuseppe Falcone, *tenenti* Ezio Massa and Italo Persegani, *s.ten.* Carlo Molinari, *m.llo* Filippo Farina, *cpr.m.* Loreto Mazzola, *caporali* Antonio Angelelli, Pietro Asteggiano and Renzo Barella; observers *ten.* Porta and *s.ten.* Degani; gunners Rinaldo Berretta, Angelo Ferrari, Lionello Mozzo and Erminio Schirio. Its aircraft were Ca 1229, 1252, 1231 and 1197.

On 29 November 1916 Caproni 1252 crashed at Aviano, killing *cap.* Falcone, who had just been appointed commander, *m.llo* Farina, *ten.* Porta and *sold.* Berretta.

Above: Ca.3 4059 crashed on the night of 4 September 1917 killing all its crew.

During the whole of 1916 the 4ª Squadriglia flew 43 combat missions.

On 8 January 1917 *cap.* Augusto Della Porta became commander, replacing *ten.* Molinari, who had been acting commander since the death of Falcone. Besides an attack on the station at San Daniele del Carso on 8 January 1917, in which an Austrian fighter repeatedly hit Ca.3 of *serg.* Barella, no actions were flown in the first months of 1917, on account of the weather. Operations resumed on 18 April 1917 with a sortie of three bombers to Chiapovano, aborted for different causes.

The first successful sortie of the year was on 14 May when Caproni 450 hp 2354 with pilots *ten.* Italo Persegani and *sold.* Francesco Acqua and gunners Schirio and Malzanini, and Ca 2369 with Della Porta, Asteggiano, Mozzo and Saletta bombed Chiapovano. On that same day *ten.* Persegani was wounded as Ca 1197 turned over on landing.

In the spring of 1917, the unit was in action supporting the 10th Battle of the Isonzo. On 25 May Ca 2374, returning from a sortie against the railroad at S. Lucia di Tolmino, crashed in flames near Indrio, its crew, formed by *cap.* Della Porta, *sold.* Acqua, gunners *cpr.* Sciaccaluga and *sold.* Mozzo, survived with minor wounds.

Its force at the time consisted of pilots *cap.* Della Porta, *tenenti* Persegani, Umberto Mongardini, Raffaele Paravicini, Carlo Toselli, *s.ten.* Vincenzo Biggio, *sergenti* Loreto Mazzola, Pietro Asteggiano, Renzo Barella, Raimondo Battistella, *cpr.* M. Luigi Remitti, *sold.* Acqua, observers *tenenti* Michelangelo Gaetani and Mario Ferrero, gunners *cpr.* Stefano Sciaccaluga and *soldati* Erminio Schirio, Angelo Ferraris, Davide Barsotti, Lionello Mozzo, Angelo Malzanini and Antonio Catozzi.

On 1 June *cap.* Della Porta passed command to *ten.* Paravicini, who was acting commander until September, when *cap.* Raffaele Tarantini replaced him. In June the squadron had Capronis 2315, 2354, 2369, 2383, in August it received the new 4059.

On 28 July 1917, the unit sent out two of its planes for the attack against the mercury processing plant at Idria. Gunner Brugo, from the back turret of Ca 2369 claimed a fighter shot down. Italian fighters were escorting the group of Capronis, and the Austrian pilot Brumowski was forced down in his damaged KD 28.57, victim either of gunner Brugo or of *ten.* Sabelli, 91ª Squadriglia. In August, besides two sorties against Pola on 8/9 August, 4ª Squadriglia flew 32 sorties in support of the Army fighting on Carso and on Hermada, most of them flown by pilots *ten.* Toselli and Paravicini, *serg.* Mazzola and Asteggiano.

Above: Ground personnel and crewmen posing at Aviano in front of a Ca.3 showing the new insignia of black lions.

In the nights of 3 and 4 September the 4ª Squadriglia took part in the large actions against Pola. On the 4th soon after take-off an engine failed Ca 4059 and the pilots tried to return to base, but the bombs went off as the plane touched down, killing the whole crew of *ten.* Carlo Toselli and *serg*. Asteggiano with gunner *cpr*. Schirio and observer *ten.* Romolo Bonafede.

On 14 September *ten.* Paravicini, *serg.* Mazzola, *serg.* Marco Resnati, gunner Malzanini and observers Toffoletto and Ferrero, were ordered to go to Taliedo to participate in the formation of Detachment A.R. for the attack against Cattaro naval base.

Like all bomber units, the 4ª Squadriglia was heavily engaged after the enemy breakthrough at Caporetto. On 25 October Ca 2383 failed to return from a bombing mission against barracks at Modrejce. Its crew of *ten.* Alfredo Trevisan, *serg.* Resnati, observer *ten.* Anacleto Carini and gunner Natale Bertone was captured. German fighters

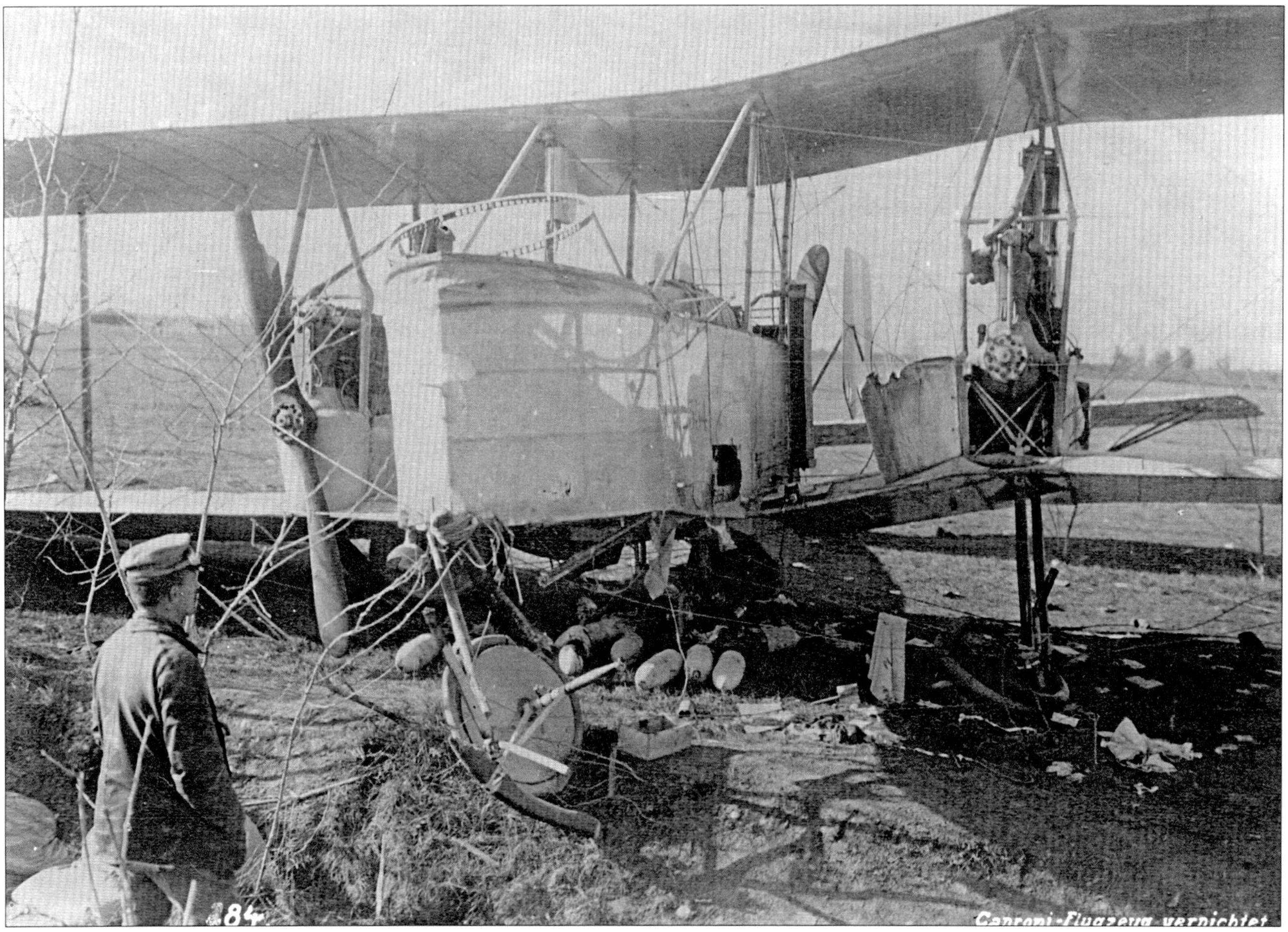

Above: Austrian postcard showing Ca.3 2354 wrecked at Campoformido after a night mission on 27 October 1917.

Above: The aircraft of the 4ª Squadriglia were identified by having one, two, three, or four lions. (Archive SHAA)

claimed four Capronis that day. Resnati together with Luigi Ridolfi, another bomber ace, crashed to his death in a Caproni loaded with journalists in August 1919. On the next day there was a night action for Ca 2354 of Tarantini, Mazzola, De Liguoro: gunner Alcibiade Borghi, on his first mission, walked on the wing to fix an engine, the damaged airplane was nursed back to Campoformido, where it crash-landed and was later burnt.

The 4ª Squadriglia with Capronis 2369, 4086, 4178 and 4185 on 1 November retreated to Padua, from where it attacked the advancing enemy. Nine bombing missions were flown in November, sometimes strafing the advancing enemy troops. On 5 December 4ª Squadriglia moved to Ghedi to be reformed with new personnel. During 1917 this unit had put up 79 combat missions.

On 28 January 1918 the 4ª Squadriglia moved from Ghedi to Verona, still part of the XI Gruppo. Its commander was still *cap.* Raffaele Tarantini, with pilots *tenenti* Ariboldo Soliani and Bernardo Viano, *serg.* Giuseppe Ponti, *caporali* Giuseppe Brambilla and Edoardo Brusadin; observers *tenenti* Francesco Caroli, Domenico Corica and Enzo De Liguoro; gunners Alcibiade Borghi, Luigi Canziani, Eligio Fontana and Fortunato Rossini. Its Capronis 450 hp had serials 4222, 4215, 4178, 4183.

The 4ª Squadriglia flew its first combat mission of 1918 on 6 February sending two aircraft, 4215 and 4222, against Mattarello at night. The latter was lost in a crash at the end of the month.

In the night of 20 February, as a Caproni had just taken off, enemy bombers arrived over the airfield of Padua, the Caproni challenged them and gunner Borghi was slighty wounded, probably by the Italian ground fire. In the night of 27 February Tarantini and Ponti with Ca 4215 bombed Mattarello again.

After a period of inaction due to the weather, bombers 4215 and 4183 attacked the same target in the night of 22 March. Anti-aircraft fire damaged Caproni 4183, with pilots *s.ten.* Gino Santi and *cpr.* Brambilla, observer *s.ten.* Nicola Verzetti and gunner *cpr.* Fortunato Rossini, the pilots decided to crash-land in the bed of the Adige River. The unlucky Verzetti was trapped in the wreck and drowned, while the other airmen were unharmed.

At this time the 4ª Squadriglia had pilots *ten.* Pirola, *s.tenenti* Gino Santi and Loreto Mazzola, *sergenti* Giuseppe Brambilla, Edoardo Brusadin and Giuseppe Ponti, *sold.* Attilio Longoni, observers *tenenti* Domenico Corica, Enzo De Liguoro, Guido Cima, Luigi Pinardi and *s.ten.* Francesco Caroli.

The power plant at Cal Sarga was attacked on 4 May. On 14 May the unit moved to Ca' degli Oppi, 6 miles south of

Above: A Caproni at Ca' degli Oppi showing the new bluish-green camouflage for night missions.

Verona. From there bombing missions against Grigno on 19 May, Fucine on 25 May were flown. On 8 June Pergine was attacked, on June 24 and 25 Conegliano, June 26 Mattarello, on June 28 only leaflets were dropped.

The new Caproni 5 bomber, six of which were sent to the front, was tested in combat, with the very expert crew of *t.col.* Armando Armani, *ten.* Luigi Ridolfi, *cap.* Natale Palotta and *sold.* Tarcisio Cantarutti temporarily operating with 4ª Squadriglia. They flew Ca.5 11585 in the action of 25 May, and the airplane performed well. They were back in action on 17 June with Ca.5 11580 landing at Villaverla, where two bombs that had remained on the plane fell on the ground, and were taken as souvenir by the British airmen of RAF No 34 Squadron. On 17 June Ca 11556 flew a mission over the Piave River; on the nights of 25 and 27 June Armani and Ridolfi took Ca 5 11580 in the first night missions. They flew again in the night of 29/30 June witnessing the loss of Caproni 4215 of *ten.* Rodolfo Calogero and *serg.* Edoardo Brusadin with observer *ten.* Francesco Caroli and gunner Canziani that caught fire and all the crew perished. An engine of Ca.5 11580 also caught fire, but the pilots made a safe landing at Verona. The Ca.5 flew two more mission on 2 and 3 July, but despite Armani's positive opinion, the pilots at the front disliked the plane, several defects were pointed out, and the Ca.5, for which a huge production program was taking place, did not go back in operations until the end of October, with 6ª Squadriglia.

At the end of July, Tarantini was assigned to the 6ª Squadriglia, and *cap.* Francesco Lanzafame became commander, after a brief period with *ten.* Giordano Bruno Pirola as acting commander. In July the unit received its first American pilots, Lieutenants Francis X. Connolly, Herbert C. Dobbs Jr., Talbot O. Freeman, Wallace Haggson, James P. Hanley Jr. and Raymond M. Robertson. The American pilots went into action dropping propaganda leaflets over enemy-held territory. In the attack of July 21 at Mattarello the only combat mission of another pilot, *ten.* Giuseppe Panzetto, is recorded.

In August the unit was using Capronis 2408, 2377, and 11503, the last two in charge to the 6ª Squadriglia. In the night of 24 August Caproni 2408 was lost, force landing after an engine failure on take-off, its crew of Lt. Dobbs and *serg.* Onofrio Tanzarella, observer *ten.* Di Anna and gunner Rossini suffering only minor injuries. Ca 450 hp 2377 and 11503, together with 4086, were formally transferred from the 6ª to the 4ª on 6 October 1918. In this period the unit was active supporting the final Italian offensive. On 26 October, 4162 crashed during a test flight, killing pilots *s.tenenti.* Giorgio Cordero di Montezemolo and Antonio Menegazzi, *sold.* Bernardo Agazzi, and *ten.* Mario Simonini of 1ª Squadriglia.

On 27 October, flying into action, Caproni 11503 with pilots *ten.* Mario Tarli and *serg.* Giannetto Vassura, gunners *soldati* Dandolo Zamboni and Domenico Fantucci, was isolated and Austrian fighters, led by *Oblt.* Roman Schmidt and Emmerich von Horvath, attacked it. The damaged bomber managed to bomb munitions dumps near Vittorio Veneto and head for home, but it could only maintain an altitude of 200 meters, easy target for enemy cannons that forced it down at Conegliano Veneto, killing the whole crew. Zamboni had been fighting the Central Empires even before Italy's entry in the war, as a volunteer with the Garibaldi Legion in France, earning the rank of sergeant and the French *Croix de Guerre* with Palms. Vassura was on leave,

The special crew that flew the first combat mission of a Caproni Ca.5 11585: at the center *t.col.* Armani, first on the left *cap*. Palotta, and last *ten*. Ridolfi.

Above: Caproni Ca.3 of 4ª Squadriglia at Ca' degli Oppi.

but he returned voluntarily to the unit as the offensive was launched, insisting to take part in the mission that cost him his life.

On 1 November the Squadriglia moved to the airfield of Padua for the final actions in support of the advancing Italians, also dropping patriotic leaflets. During the year 1918 the 4ª Squadriglia, which marked its aircraft with black lions as insignia, had flown 71 combat missions, the total for the whole war was 216, paid with 26 casualties. Lt. Dobbs had flown six missions, Lt. Freeman two, Lt. Hogson, Lt. Hanley, Lt. Robertson one each. By the new year, under the command of *cap.* Cellerino it was equipped with eleven Ca 600 hp with Fiat A12 engines. This unit was disbanded at Campoformido in the late summer 1919.

Casualties and Combat Losses of the 4ª Squadriglia

Date	Place	Aircraft	Crew	Cause
18.2.1916	Merna	Ca 703	*Cap.* Visconti KIA, *cap.* Turilli POW	Fighters
25.6.1916	Bassano	Ca 704	*Cap.* Ungania KIA, *s.ten.* Mazzetti KIA. *sold.* Visioli KIA	Fighters
1.8.1916	Fiume	Ca 1228	*S.ten.* Signorini POW, *cpr.* Caparello KIA. *cpr.* Sbaraglia POW	Fighters
6.8.1916	Miramare	Ca 1134	*S.ten.* Zanini KIA, *serg.* Borghi WIA/POW. *cpr.* Manieri WIA/POW	Fighters
29.11.1916	Aviano	Ca 1252?	*Cap.* Falcone KIC, *m.llo* Farina KIC. *ten.* Porta KIC, *sold.* Beretta KIC	Accident
25.5.1917	Indrio	Ca 2374	*Cap.* Della Porta UNH, *sold.* Acqua UNH, *cpr.* Sciaccaluga UNH, *sold.* Mozzo UNH	Ground fire
4.9.1917	Aviano	Ca 4059	*Ten.* Toselli KIC, *cpr.* Asteggiano KIC, *ten.* Bonafede KIC, *cpr.* Schirio KIC	Accident
25.10.1917	Carso	Ca 2383	*Ten.* Trevisan POW, *serg.* Resnati POW, WIA/POW, Bertonc POW *ten.* Carini	Fighters
22.3.1918	Borghetto	Ca 4183	*S.ten.* Santi UNH, *cpr.* Brambilla UNH, *s.ten.* Verzetti KIA, *cpr.* Rossini UNH	Ground fire
30.6.1918	Near Verona	Ca 4215	*Ten.* Calogero KIC, *serg.* Brusadin KIC, *ten.* Caroli KIC, *sold.* Canziani KIC	Accident
26.10.1918	Ca' degli Oppi	Ca 4162	*Ten.* Simonini KIC, *ten.* Menegazzi KIC, *Ten.* Cordero di M. KIC, *sold.* Agazzi KIC	Accident
27.10.1918	Conegliano	Ca 11503	*Ten.* Tarli KIA, *serg.* Vassura KIA, *sold.* Fantucci KIA, *sold.* Zamboni KIA	Fighters & ground fire

Above: The crash of Ca.3 4183 in the Adige River on 22 March 1918 in which *s.ten*. Verzetti lost his life.

Above: Ca.5 11585 on 17 June, after landing at Villaverla, British personnel greet the Italian crew, and on the ground there are the bombs that fell on the airfield. The British, northern Italy being considered part of the Near East, are wearing pith helmets. (Archive Caliaro)

Above: Ca3 11503 passed from 6ª Squadriglia to 4ª Squadriglia in the last days of the war. Its Latin motto is "*Unus non sufficit orbis*" means "One world is not enough". *Serg.* Vassura is the first on the right. (Archive Caliaro)

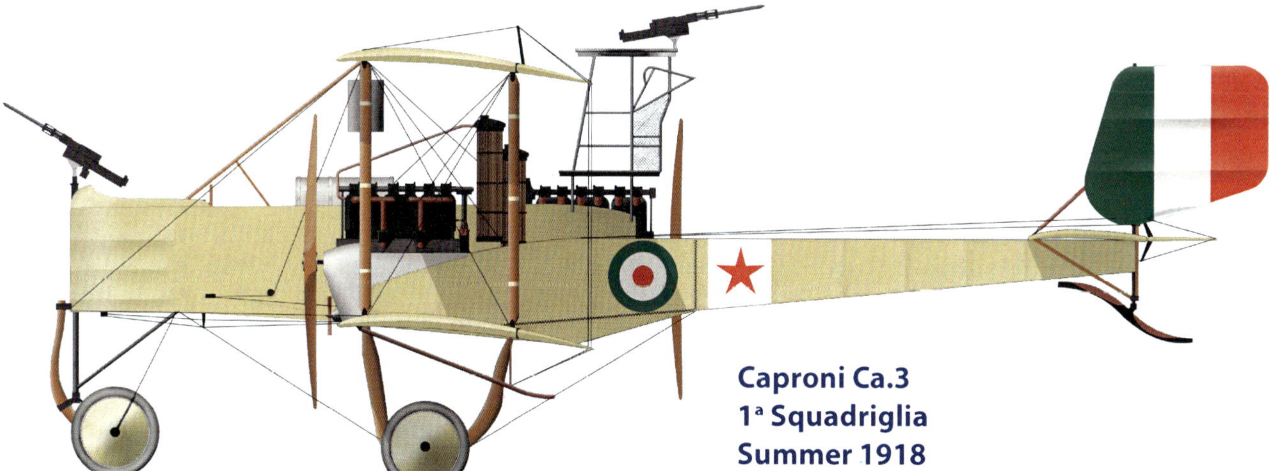

Caproni Ca.3
1ª Squadriglia
Summer 1918

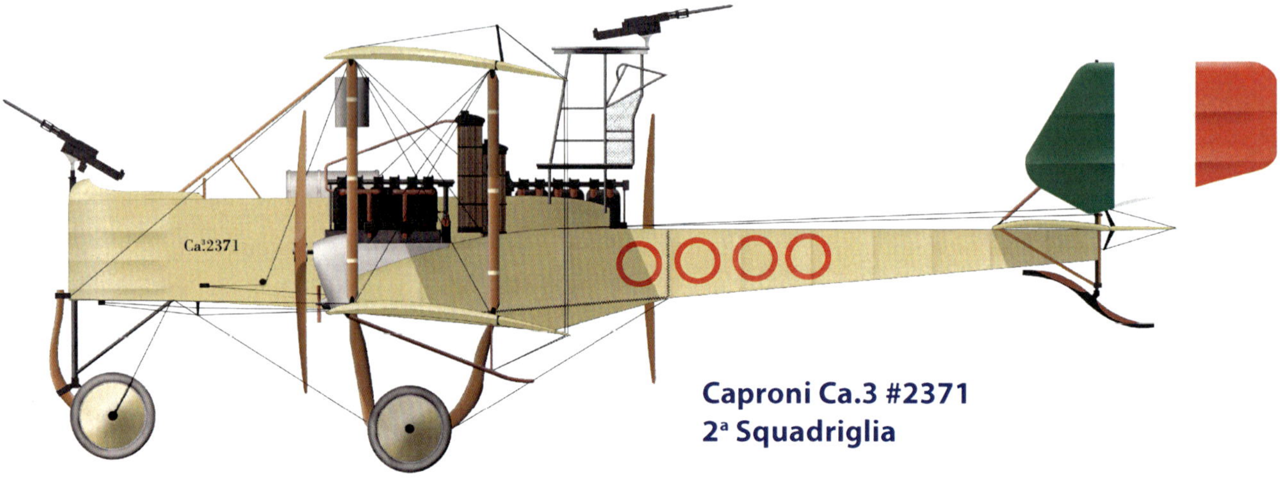

Caproni Ca.3 #2371
2ª Squadriglia

Caproni Ca.1 #1136
3ª Squadriglia
Ca. 1916

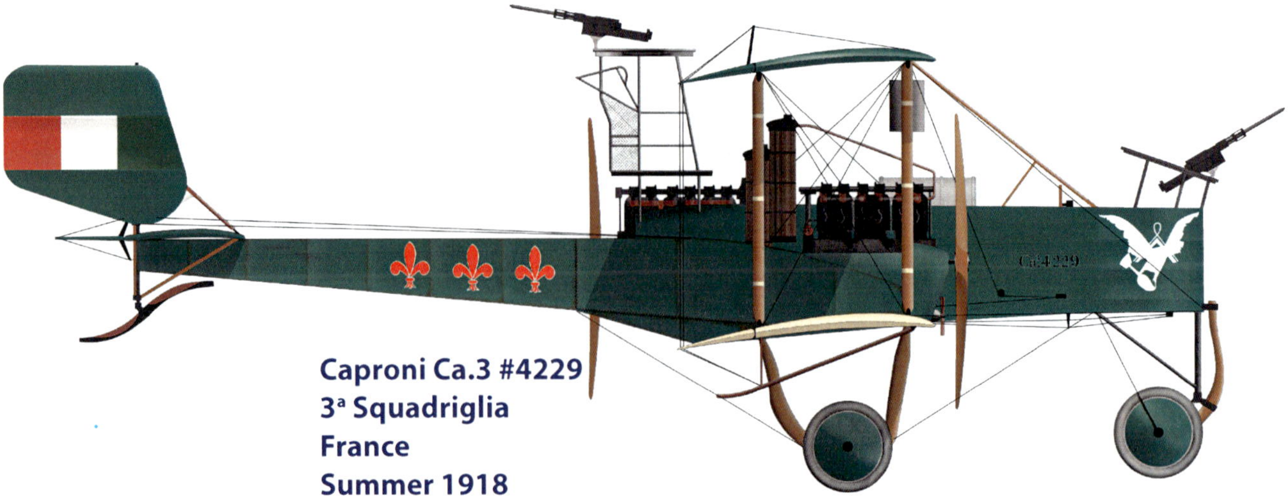

Caproni Ca.3 #4229
3ª Squadriglia
France
Summer 1918

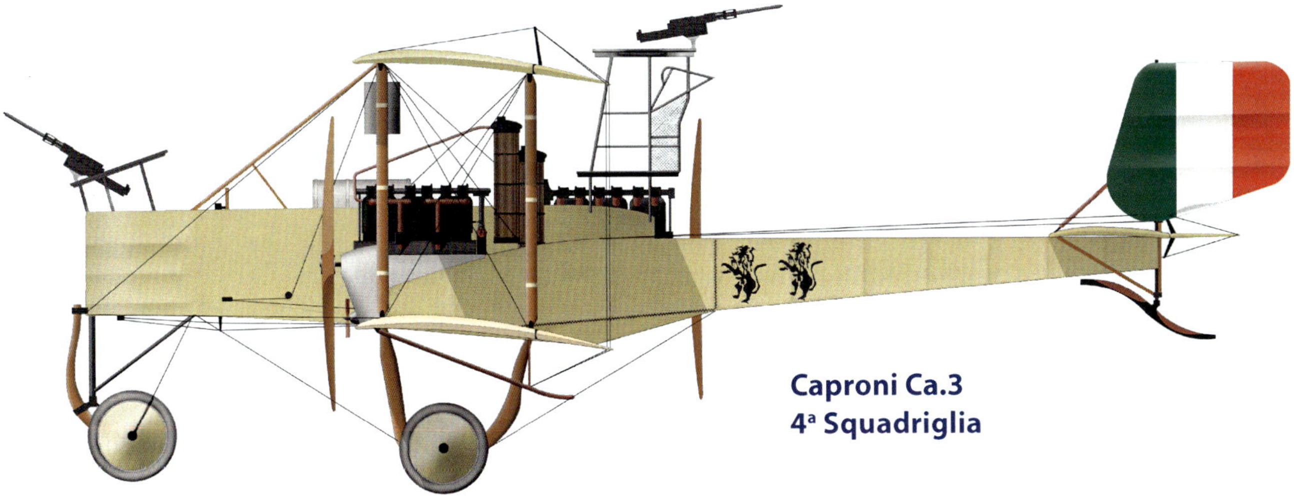

Caproni Ca.3
4ª Squadriglia

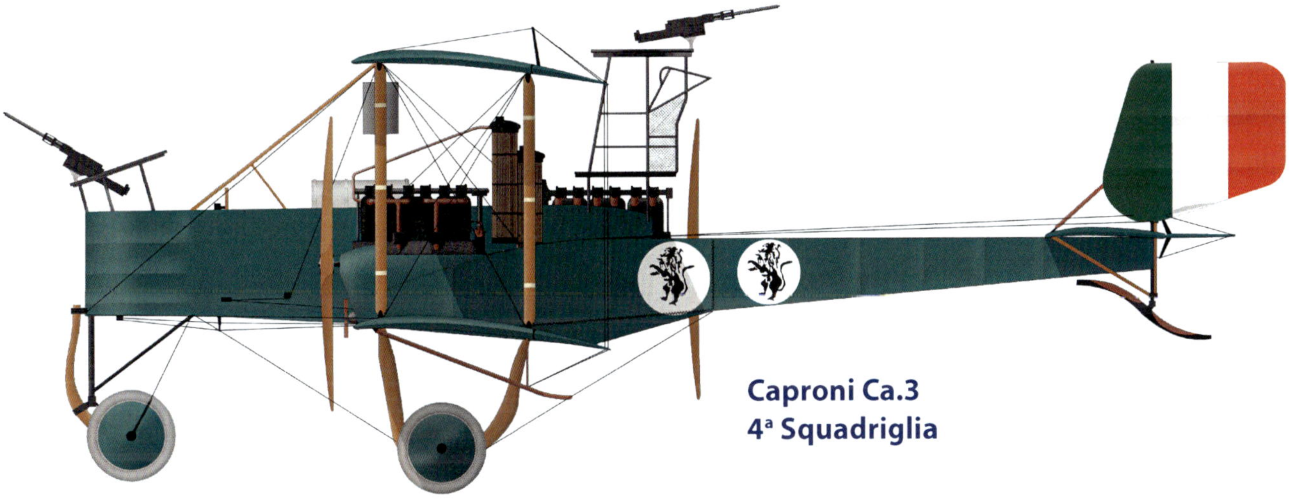

Caproni Ca.3
4ª Squadriglia

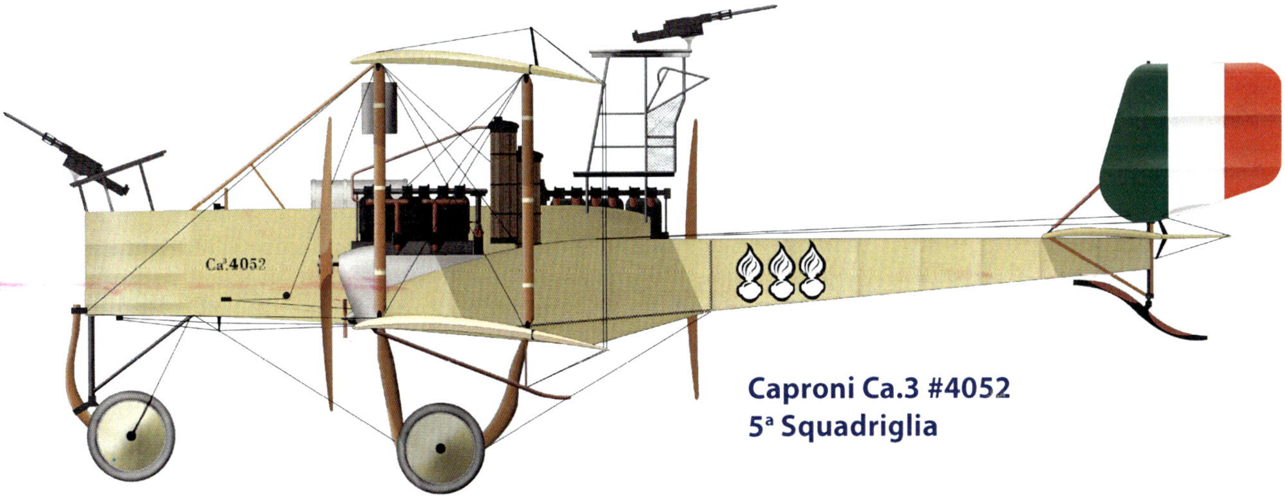

Caproni Ca.3 #4052
5ª Squadriglia

5ª Squadriglia

Above: The insignia of the 5ª Squadriglia.

Above: Milan Stefanik, wearing the decorations of Officier de la Legion d'Honneur, the Croix de Guerre with Palms, and the Italian Ordine dei SS. Maurizio e Lazzaro.

The 5ª Squadriglia was formed on 22 November 1915 at Aviano, and on 16 December it moved to Tombetta field, Verona, assigned to the III Gruppo operating for the 1st Army, on the north-eastern part of the front. It was not, therefore, part of the strategic force of the Supreme Command. It began operations on 10 January 1916 bombing the Austrian airfield at Gardolo. The unit's commander was *cap. pilota* Oronzo Andriani, its pilots were *capitani* Gaetano Coniglio and Ettore Croce and *serg.* Igino Bevilacqua, observers *capitani* Enzo Baroni and Pietro Bentivegna.

The Squadriglia was engaged in bombing and reconnaissance flights and it even flew standing patrols when enemy aircraft were reported approaching the lines. On 27 March 1916 two Capronis took off from Verona to protect the city from raiders, and Ca 714 did indeed intercept an Austrian airplane and forced it to retreat.

On 16 April, according to the new organization, this unit assumed the name of 5ª Squadriglia Aeroplani. On 4 May *cap.* Andriani flew with Lieut. Milan Stefanik, of the French Army, actually the General who was the herald of the independence of Slovakia, dropping messages and repeated this action on 7 May during a bombing mission to Mattarello.

On 30 May 1916 the commander was *cap.* Coniglio, with pilots *s.ten.* Vittorio Bianchi, *serg.m.* Stefano Caselli, *sergenti* Bevilacqua, Riccardo De Paolini and Pietro Ferri, observers *cap.* Pietro Bentivegna and *ten.* Giulio Franciosini. Equipment included Capronis Ca 1155, 1144 and 1211.

On 2 June, as the Austrians had launched their "Strafexpedition" offensive, the 5ª Squadriglia claimed a victory in air combat as three airplanes were bombing Calliano, and Ca 1155 with pilots *serg.* De Paolini and Ferri, gunner Emilio Blesso and observer *ten.* Franciosini, faced three Austrian planes. A biplane identified as "type Lloyd" was seen leaving the combat in a steep dive with its engine stopped, possibly Lohner B.VII 17.39 of Flik 17, that force landed at Gardolo with a wounded pilot. On 8 June Ca 1155 itself fell victim of anti-aircraft guns during a mission against the Rovereto railway station, it crash-landed with no consequences for its crew.

In July 1916 the 5ª Squadriglia was still at Tombetta di Verona in the III Gruppo with commander *cap.* Coniglio. Its pilots then were *s.ten.* Bianchi, *serg.m.* Caselli, *sergenti* Bevilacqua, Ferri, Vittorio Romagnino, *cpr.* Guido Sobrero, *soldati* Antonio Ferrari and Mario Zoppola; observers *capitani* Bentivegna, Franciosini, Guido Barutta and Mario Ajmone Cat, *tenenti* Franciosini, Federico Caneva (the son of the General commander of the Italian troops in the Libyan war of 1911) and Giuseppe Cipriano, *s.ten.* Giovanni Cartia and Dino Dani; gunners Raffaele Antonucci, Romeo

Above: Caproni Ca.1 714 was the first airplane of 5ª Squadriglia.

Betteghella, Enea Mariuzzo and Carlo Vitrotti. It had on strength Ca 1141, 1180, 1204, 1211, 1242. In the summer of 1916, the 5ª Squadriglia was active supporting the operations leading to the conquest of Gorizia, the 6th Battle of the Isonzo, as well as in its Trentino area of operations.

On 23 July Caproni 1173, that had reached the unit just six days before, crashed for unknown reasons right after take-off, killing its pilots *s.ten.* Bianchi and *serg.* Ferri and the observer *cap.* Bentivegna, while gunner Antonucci was injured. On 23 August another "Lloyd" was claimed as shot down as it tried to attack Ca 1177 of Bevilacqua, Sobrero, Cartia and Mariuzzo above Arsiero. Two days later, 25 August 1916, the squadron suffered its first loss due to enemy air action, as Ca 1179 with Coniglio, Sobrero, Dani and Vitrotti failed to return from an action against the railroad station of S. Cristofaro, near the Caldonazzo Lake. The winner was Brandenburg C.1 serial 26.29 piloted by the future ace Josef Kiss with the observer Lt. Fiedler, that received 70 hits from the machine gun of the Caproni. The Italian bomber crash landed near the fort of Luserna, with *cap.* Coniglio, a very active and much-loved commander, dead and Sobrero and Vitrotti wounded. *Ten.* Franciosini became acting commander until 30 August, when *cap. pilota* Filippo Valdimiro replaced him. New airmen arrived: the pilots *tenenti* Mario Giulio Agostini, Max Arici, Ettore Della Cella, *cpr.* Emilio Lodesani and the observer *ten.* G. Battista Lucci Chiarissi.

On 17 September 1916 Ca 1174 of Valdimiro, Bevilacqua and Blesso was forced down and crash landed in no man's land near Ghizzola. The bomber was later recovered but Valdimiro died of his wounds six days later. Thus, the 5ª Squadriglia had lost two commanders in less than one month. It was Kiss again who had shot down the bomber, flying a Hansa-Brandeburg C.1 with observer *Oberleutnant* Karl Keizar.

Cap. Umberto Rossi took over command of the squadron on 5 October. In December 1916 the 5ª Squadriglia had on strength Ca 1192, 1226, 2605, 1240, 1172 and 1177. Its personnel included pilots *cap.* Rossi, commander, *ten.* Agostini, Arici and Della Cella, *s.ten.* Ferrari and Giuseppe De Agostinis, *serg.* Bevilacqua and Vittorio Romagnino, *cpr.* Lodesani, *sold.* Corrado Avenati; observers *cap.* Barutta, *ten.* Caneva and Franciosini, *s.ten.* Vincenzo Bonfiglio; gunners Betteghella, Blesso and Vitrotti. Mario Giulio Agostini, who was an actor, wrote excellent and honest commentaries after the war, including a book, '*Ali della Strage*' (Wings of the slaughter) before his untimely death in a road accident in 1919.

In the course of 1916, the 5ª Squadriglia flew 45 bombing missions, 45 standing patrols and 6 reconnaissance flights and had 12 air combats. On 26 February 1917 Ca 1174 was used for experiments in wireless telegraphy.

On 10 March 1917 ground fire pierced the fuel tanks of Ca 1226 and its engines stopped. Pilots Agostini and Lodesani with gunner Gino Salvadori completed the mission bombing Malga Cheserle, then glided to a crash landing near the Adige River, the observer *ten.* Lucci Chiarissi suffered a broken leg. The action was flown at low level on account of the weather; the Caproni of Arici and Ferrari was also hit. On 26 March Ca 1175 of *s.ten.* Francesco Notarbartolo and *serg.* Bevilacqua, with *s.ten.* Bonfiglio and Vitrotti, claimed a Brandenburg that had attacked over Col Santo and was seen

Above: Ca.3 1173 that crashed on 23 July 1916 with the death of three of its crew.

going down out of control.

On 23 May the 5ª Squadriglia, together with the 9ª, moved to La Comina airfield to support the 10th offensive of the Isonzo. On 24 May Ca 2362 of *ten.* Della Cella, *serg.* Romagnino, Salvadori and Blesso, flying in the afternoon to attack Austrian positions on the Hermada hill, faced the attack of two enemy fighters. The able gunner Blesso, already decorated with a Silver Medal for the action of 17 September 1916, shot down one of the attackers, possibly KD 28.25 of Flik 42, with Lt. Melzer wounded, but was himself mortally wounded. The bomber made an emergency landing near Latisana and was written off.

On 1 June 1917 the 5ª Squadriglia returned to Tombetta di Verona. The next day *s.ten.* Notarbartolo was killed in a flying accident when Ca 1149 crashed near Taliedo. This officer was attached to the Comando Aeronautica since 24 March. On 10 June Ca 2368, after bombing in Val Portule, crashed, probably victim of Austrian fighters. Nobody survived from its crew of *tenenti* Arici, *s.ten.* Lodesani, Federico Caneva and the veteran gunner *m.llo* Betteghella. Caneva by that time was assigned to 31ª Squadriglia, but he had asked to go on the mission with his old unit.

On 19 June 1917 during a bombing mission in Val Galmarara of Capronis 2365 and 2384, the former was forced back by failing engines, the latter dropped 12 bombs and, on the way back, gunner Vitrotti was forced to walk on the wing to fix a leaking water pipe of the right engine.

On 6 August the unit was moved to Campoformido and was assigned to the IV Gruppo: four days later it got a new commander, *cap.* Umberto Bassani. At that date it had Capronis 2384, 2377, 2365 and 2356, and pilots *tenenti* Agostini, Della Cella, Ferrari, Pietro Gallia and Giuseppe Orselli, *sergenti* Bevilacqua and Romagnino; observers *cap.* Natale Palotta, *ten.* Fausto Da Passano, *ten.* Renzo Lanfranco; gunners Luigi Griffini, Giovanbattista Mariano, Gino Salvadori and Enrico Vanzulli.

On 19 August 1917 5ª Squadriglia went into action during the 11th Battle of the Isonzo. Three bombers, 2356, 2384 and 4052, bombed enemy positions at Mahorini Vojscica and Cerovlje, dropping bombs from an altitude of 10,000 feet and then going down to 3,000 feet for strafing passes. Ca 2356 of *ten.* Gallia, Orselli, Da Passano and Mariano, with its central propeller damaged by anti-aircraft fire, faced the attack of two fighters, one being KD 28.38 of Flik 12 with pilot Benno Fiala. The Austrians didn't press their attack and the Italian bomber, flying on a single engine, managed a forced landing, smashing its landing gear and wingtips, but bringing its crew home unharmed. The unit was active in the offensive all through the month, hitting enemy troops on the back of the Hermada, at Grahovo and in the valley of Chiapovano, the limit of the Italian penetration. In the night of 3/4 September the 5ª Squadriglia flew its first

Right: Mario Giulio Agostini at the controls of Caproni 1175. In this photo the Type E bombing sight can be clearly seen. (Archive Fiorenzo Longhi)

night mission bombing Pola, *ten.* Gallia had to face again Austrian seaplane fighters with his Caproni already damaged by ground fire.

On 20 September Agostini, Bevilacqua, Da Passano Vanzulli and engineer Giaime drew lots and went to Taliedo, to be part of the Distaccamento A.R. for the action against Cattaro naval base.

On 27 September two Capronis took off at 21.45 p.m., headed for Pola. They were 4061, with crew of Bassani, Gallia, Lanfranco and *serg.* gunner Giovambattista Moriano, and 4171 with Della Cella, Ferrari, Palotta and Salvadori. The bombers dropped two 260-mm and six 162-mm bombs over the harbor, the arsenal and the moored warships, then on the return flight they both got lost in the haze, and 4171, running out of fuel, barely managed to land at Arcade airfield. The next day, 28 September, 4070 and 4061 were tasked with a mission against positions at Cerovlje and Vojscica. As it was dropping its bombs, 4070 was hit by a shell that ripped through its wing. Della Cella and Ferrari completed their action and nursed the bomber back to the Italian lines, and while landing in the parade grounds of Gorizia, finding a bomb crater where the bomber suffered further damage. On 29 September there was another night attack against Pola.

Above: A Nieuport XI, 1628, is posed next to Ca 1169 of 5ª Squadriglia, possibly at Tombetta airfield.

Above: Shot down on 25 August 1916 with the death of squadron commander *cap*. Coniglio, Caproni 300 hp 1179 draws the attention of Austrian soldiers.

On 5 October the Comando Raggruppamento ordered the 5ª Squadriglia to move to Aviano with its Ca 4140, 4061, 4174 and 4165, with commander Bassani and pilots Della Cella, Gallia, Ferrari, Orselli, Romagnino and *serg.* Candido Cagliero; observers *ten.* Leonardo Mariani, Palotta and *s.ten.* Raul Gattermayer; gunners Griffini, Salvadori and Bomprino. The unit was now assigned to the XI Gruppo Aeroplani. A winged grenade was the insignia of this unit. On 17 October the personnel that had been sent to Gioia del Colle rejoined the unit.

During the Caporetto retreat, the 5ª Squadriglia flew constantly. On 26 October Ca 4147 with crew Della Cella, Romagnino, gunners *serg.m.* Federico Bissacco and *sold.* Giuseppe Tolentini failed to return from a mission over S. Lucia di Tolmino against the advancing Germans. Its fate was only known after the war: ten German fighters had forced it down with Romagnino fatally wounded (he died ten days later). Bissacco managed to escape from prison camp, but Della Cella returned from captivity in Germany after the end of the war: he was discovered by the French wandering, completely crazy, his body showing the scars of tortures. He died in 1919 without ever having told his tragic story.

On 31 October the 5ª Squadriglia left Aviano and moved to Padua with its Caproni 4165, 4061, 4140 and 4189. Equipment at Aviano was destroyed to prevent its capture by the Austrians. Caproni 2384 was set on fire, as it was unfit to fly. The mechanics of the squadron managed to recover its instruments. On 1 November Ca 4061 and 4189 were sent to La Comina to pick up canned beef to be dropped to Italian troops at Codroipo. While at La Comina, Ca 4189 was damaged in an enemy air attack. In the days of the retreat the squadron bombed by day and night the advancing enemy troops, particularly hitting the bridges of Pinzano al Tagliamento and over the Livenza River and the Codroipo-Udine road. Having dropped their bombs, the Capronis often also strafed the troops harassing the routed 2ª Armata. On 13 November they hit the Silesian 12th Division that tried to cross the Piave at Vidor. As the offensive petered out in the South, the Austrians tried to break through in the sector of the Plateaus and the Grappa Mountain, heading for the valleys that would allow them to encircle the Italian defenses along the Piave. The 5ª Squadriglia therefore struck at the communication lines in that mountain area. On 27 November Ca 4184, with *cap.* Bassani, *cap.* Gallia, *ten.* Mariani and *sold.* Griffini, and 4189 with *ten.* Manlio Borri, *m.llo* Goffredo Corso, *s.ten.* Gattermayer and *sold.* Giuseppe Garzena took off to bomb Primolano. Fighters attacked both of them on their way back, and Garzena was wounded, so

Above: The details of the loss of Caproni 1177 of 5ª Squadriglia are not known.

Caproni 4184 had to face two enemies, but one was driven down and the other one gave up, after a twenty-minute combat.

On 8 December the 5ª Squadriglia moved to Ghedi with its Ca 4184, 4228 and 4165. The last one got lost in the dark and made an emergency landing smashing its wings. In January 1918 the unit, flying Capronis 4184 and 4228 and with commander *cap.* Bassani, had pilots *tenenti* Bevilacqua, Borri, Ferrari, Gallia, Orselli, *serg.* Davide Dall'Aglio, observers *tenenti* Fantoni Scotti and Leonardo Mariani, and gunner *sold.* Bomprino. On 3 February it was reassigned to the IV Gruppo and moved from Ghedi to San Pelagio, where two days later it received Ca 4081 from Taliedo. During February, as weather conditions were bad, the unit prepared its aircraft and personnel, then it flew by night against the stations at Bolzano on 27 February and Primolano, with three planes, on 29 March.

On 4 May Ca 4081 turned over in take-off. Ferrari and Bevilacqua, after a long service, were retired from the front. The 5ª Squadriglia now included *cap.* Gallia, *tenenti* Agostini, Orselli, *m.llo* Goffredo Corso, *sergenti* Davide Cagliero, Dall'Aglio, Romeo Parietti, Umberto Pansiera, observers *tenenti* Leonardo Mariani, Zeno Verga, Giuseppe Belsito, Raul Gattermayer, Mario Pansa, Pietro Scotti Fantoni, gunners *sergenti* Sergio Buri and Gino Raimondi, *cpr.* Livio Ambrosioni, *soldati* Enrico Vanzulli, Luigi Griffini, Giovanni Mai, Parmenide Massimi, Garzena.

On 4 May all its planes bombed the power plant at Fiesso. On 14 May the 5ª Squadriglia was tasked to bomb Santa Giustina airfield, in a formation of ten Caproni, the idea being that the Austrian fighters would take off, to be bounced by twenty fighters from Quinto di Treviso. The bombing was effective but the trap did not spring, as the Austrians remained on the ground. The enemy air force returned in strength on 19 May, attacking the Capronis as they were bombing marshalling yards at Osteria Palù, but losing three of their numbers, of Flik 14, to Italian and British fighters. The Caproni of *ten.* Agostini and *serg.* Cagliero was badly damaged but completed its mission. On 24 May three crews completed the effective bombing of Feltre airfield. Attacks on Austrian airfields continued: Belluno on 1 June, Motta di Livenza on 6 June. On 9 June two planes took part in the bombing of the station at Pramaggiore.

In mid-June the Austrians launched their final offensive along 130 km of front. On 15 June Ca 11492 and 11496 bombed troops that were fording the Piave using boat bridges, and then 11496 went down to strafe. On 18 June,

Above: Caproni 4052 showing the grenade markings of 5ª Squadriglia. (Collection Dino Uliana)

returning from a mission over Ponte di Piave, already damaged by ground fire, 11496 broke its landing gear in an emergency landing at Marcon, but it was back in action two days later. The next day the unit received from the IV Gruppo its first Caproni Ca.5 serial 11537, which however was used only for test and familiarization flights, as all units using this type couldn't overcome its teething troubles. On 20 June the unit got its first American pilots and, on that day, Lieuts. Pilots Donald G. Frost and Ira W. Galehouse took part in a bombing of Falzè, the king of Italy immediately awarded them the War Cross. Their politeness, courage and capacity were much appreciated.

On 17 July two planes of 5ª Squadriglia took part in the large daylight raid of 16 Capronis against Pola. Ca 11496 of Orselli, Corso, Gattermeyer and gunner Giovanni Mai hit Scoglio Ulivi and the arsenal, Ca 11492 of Agostini, Frost, observer *ten.* Mario Pansa and Vanzulli attacked the arsenal. On 18 July Ca 11492 returning from a leaflet dropping mission met a strong tailwind while landing and ended upside down in a pond at the end of the runway. Gunner *serg.* Pietro Buri, an irredentist, was trapped in the wreck, he was promptly rescued but he died the next day. Pilot Gallia, Lieut. Warren G. Wheeler and the observer *ten.* Pietro Scotti Fantoni survived.

On 23 July, also returning from a leaflet mission, Ca 11496 too had an accident: its engines failed on landing and the bomber spun into the ground. Its pilot *m.llo* Goffredo Corso was fatally injured, the last casualty of a unit that had so many losses that it was nicknamed *"la Squadriglia della Morte"* (death squadron). Pilot *serg.* Cagliero, *ten.* observer Zeno Verga and gunner Gino Raimondi were wounded.

In August the 5ª Squadriglia, at the command of *cap.* Bassani, had pilots Lieuts. Frost, Galehouse and Wheeler, *tenenti* Agostini, Gallia, Orselli, *sergenti* Dall'Aglio, Umberto Panziera, Silvio Geri, Romeo Parietti and Pietro Nave; observers *tenenti* Gattermayer, Pansa and Belsito; gunners Garzena, Griffini, Vanzulli and Silvio Ambrogioni and the Caproni Ca.3 4079, 4081, 4228 and Ca. 5 11537.

On 23 August the airfields of Aviano and Comina were bombed again. The action was repeated on 24 August and four Austrian fighters attacked Ca 4081, which managed to shoot one of them down, then escorting Spads drew away the other ones. On 31 August, Lieut. Wheeler was called back to the USA as a test pilot, Lieut. Aubrey Russel replaced him. On 1 September again the unit attacked Aviano with two bombers, pilots cap. Gallia and Lieut. Russel, ten. Orselli and Maj. Fiorello La Guardia, member of the U.S. Congress and future Mayor of New York.

Above: Ca.3 11492, two Fiat machine guns are fitted on the back turret.

On 14 September La Guardia flew again with the 5ª Squadriglia, participating in a night mission against Pergine airfield. He also flew in a night mission dropping leaflets on the 15th. Other night missions struck the airfield of Godega di Sant'Urbano on 16 September and Mansuè on 17 September. On 20 September Ca 4228 and 4081 were parked at Padua for a big display for King Victor Emmanuel III.

In October *cap.* Pietro Gallia assumed command of this unit. Weather precluded flying until the 23rd, when Ca 4228 and 4081 bombed tents near Feltre, in the preparation for the final Italian offensive. This was Agostini's fiftieth combat mission and he was the first pilot of the squadron to reach this mark. On 25 October Ca 4172 and 4081 attacked Primolano, Ca 4172 hit a freight train near the station of Grigno. On 27 October two Capronis bombed in the area of Livenza River. The final mission of the war was on 31 October, when Ca 4228 and 4081 bombed the road from Sacile to Pordenone.

At the end of the war the 5ª Squadriglia had Capronis Ca 4081, 4228, 4172, 4079 and Ca.5 11537. Its personnel consisted of commander *cap.* Gallia, pilots Lieuts. Frost, Galehouse, Russel, *tenenti* Agostini and Primo Osti, and *serg.* Geri, Nave and Alfonso Coscarella; observers Gattermayer, Mariani, Pansa and *cap.* Giuseppe De Leva; gunners Ambrogioni, Vanzulli, Guglielmo Fini, Massimi and Andrea Lovera.

On 10 November *cap.* Gallia flew Caproni 4081 over Trieste with Lt. Russel, Mariani and Ambrogioni to take photographs of the city. On 2 March 1919 Ca 4172 flew to Vienna with pilots *ten.* Mario Giuliani Gusman and *serg.* Parietti, observer *cap.* Antonio Landini and engineer *cpr.* Cesare Maroni.

The 5ª Squadriglia was disbanded in the late summer 1919.

Casualties and Combat Losses of the 5ª Squadriglia

Date	Place	Aircraft	Crew	Cause
8.6.1916	Veneto	Ca 1155	*S.ten.* Bianchi UNH, *serg.m.* Stratta UNH *cap.* Barutta UNH, *sold.* Nascimbeni UNH	Ground fire
23.7.1916	Verona	Ca 1173	*S.ten.* Bianchi KIC, *serg.* Ferri KIC *cap.* Bentivegna KIC, Antonucci WIC	Accident
25.8.1916	Luserna	Ca 1179	*Cap.* Coniglio KIA, *s.ten.* Dani POW *cpr.* Sobrero POW, *sold.* Vitrotti POW	Fighters
17.9.1916	Adige	Ca 1174	*Cap.* Valdimiro KIA, *serg.*Bevilacqua WIA, *sold.* Blesso WIA	Fighters
10.3.1917	Malga Chiserle	Ca 1226	*Ten.* Agostini UNH, *cpr.* Lodesani UNH *ten.* Lucci WIA, *sold.* Salvadori UNH	Ground fire
24.5.1917	Selo	Ca 2362	*Ten.* Della Cella, *serg.* Romagnino, *sold.* Salvadori, *sold.* Blesso KIA	Fighters
2.6.1917	Taliedo	Ca 1149	*S.ten.* Notarbartolo KIC, *serg.* Boero WIC, Sacconi WIC	Accident
10.6.1917	Val d'Astico	Ca 2368	*Ten.* Arici KIA, *ten.* Lodesani KIA *ten.* Caneva KIA, *m.llo* Betteghella KIA	Fighters
19.8.1917	Faiti	Ca.2356	*Ten.* Gallia UNH, *ten.* Orselli Unh *ten.* Da Passano UNH, *serg.* Moriano UNH	Fighters
26.10.1917	S. Lucia	Ca 4147	*Ten.* Della Cella POW, Bissacco POW *sold.* Tolentini POW, *serg.* Romagnino KIA	Fighters
27.11.1917	Primolano	Ca 4189	*Ten.* Borri UNH, *s.ten.* Gattermayer UNH *m.llo* Corso UNH. *sold.*Garzena WIA	Fighters
18.7.1918	San Pelagio	Ca 11492	*Ten.* Gallia UNH, Lt. Wheeler UNH *ten.* Scotti Fantoni UNH, *serg.* Buri KIC	Accident
23.7.1918	Mezzano	Ca 11496	*Serg.* Cagliero WIC, *m.llo* Corso KIC *ten.* Verga WIC, *serg.* Raimondi WIC	Accident

American pilot Fiorello La Guardia, at left, later became mayor of New York City. With him are *magg.* Pietro Negrotto, a senator, *cap.* Zapelloni, and gunner Otello Firmani.

Above: A marble memorial slab at Vizzola, formerly a Caproni factory and now the Volandia aviation museum, showing the raids of Caproni bombers in France.

Caproni 703, shot down on 18 February 1916, was much photographed for propaganda, the Austrians then recovered it and tested it with serial 00.25.

6ª Squadriglia

Above: The insignia of the 6ª Squadriglia.

Above: An Italian flag is painted on the central nacelle of Caproni 1231.

The 6ª Squadriglia was formed at Pordenone on 11 November 1915, assigned to the Comando del Battaglione Squadriglie Aviatori. and its commander was *cap. pilota* Adolfo Resio. It had on strength two Capronis and a third one as reserve, among them Ca 711. Besides the commander, there were *cap. pil.* Piero Gregorini, *serg. pil.* Federico Minardi and *Ten. di Vascello* Paolo Vandone, a naval officer who died in 1925 as commander of submarine *Veniero*. On 1 December it was named 6ª Squadriglia Offesa. It reached Aviano on 18 December.

In the end of 1915 and in early 1916 this Squadriglia was engaged in reconnaissance, bombing and standing patrol missions. On 6 January 1916 Ca 711 flew in reconnaissance to Fiume. On 1 March 1916 Resio was shifted to the 3ª Squadriglia and *cap.* Ferruccio Rossi, a doctor, replaced him on the 4th, that same day the unit received also *s.ten.* Edoardo Scavini.

First mission of the year was flown on 31 March 1916, a reconnaissance over Adelsberg by Ca 479, borrowed from the 1ª Squadriglia together with other Capronis of the 7ª. The crew, Rossi and Minardi with the observer *s.ten.* Mario Olivieri and gunner Oreste Cornalba, wore armors as a protection against enemy fire. Indeed, over Muggia a Fokker made an unsuccessful attack on the Italian airplane. The same crew with gunner *sold.* Aldo Buratti dropped bombs at Ranziano.

On 16 April, now named 6ª Squadriglia, based at Aviano, it was assigned to the new IV Gruppo. At that time, it had commander *cap.* Rossi, pilots *serg.* Minardi, *serg.* Lodovico Montegani and *s.ten.* Scavini; observer Olivieri and gunners Aldo Buratti and Alberto De Rossi. It had on strength Capronis 711, 1157 and 1142.

On 20 April the seaplane station of Trieste was attacked. Ca 1142 with the crew Scavini, *serg.* Montegani and observer Olivieri, completed the mission with only two engines running.

On 1 May Ca 1157 attacked Ranziano. On 16 May Scavini and Montegani flew Ca 711 to bomb Lokvica. On 30 May the two pilots bombed Mount Erio with Ca 1142 together with Rossi and Minardi on Ca 711, they repeated the mission on 31 May.

Above: Caproni 711 over the tracks of the little carts used to move the airplanes out of the hangar at Aviano. (Collection Ferruccio Ballista)

In May and June 1916, despite unfavorable weather, the 6ª Squadriglia supported the 1ª Armata in its effort to contain the Austrian offensive from the North, bombing objectives near Monte Erio and other locations of the Asiago Plateau. On 30 May Ca 711 suffered a fuel system failure and crash landed near Godega di Sant'Urbano, without harm for its crew of Minardi, Rossi and Olivieri. On 20 June Ca 1142 attacked the enemy airfield at Ciré and then with its defensive fire kept at bay four Fokkers that attacked it one after the other. On 7 July Rossi and Minardi with observer Olivieri flew 1157 to bomb troops at Ghertele and managed to bring back the plane although badly damaged by ground fire.

On 1 August 1916 the unit took part in a mass bombing of the Whitehead torpedo factory at Fiume, sending all its serviceable aircraft. These were Ca 1157 of Rossi, Minardi and Olivieri, 1142 of Scavini, Montegani and gunner Guido Ciastellardi, and 1219 of pilots *s.ten.* Umberto Mongardini and *serg.* Cesare Baccili and gunner Riccardo Picco. In the afternoon of 9 August, the same bombers hit Prevcina, as the 6th Battle of the Isonzo was going on. This time the crew of Ca 1157 was pilots *s.ten.* Ignazio Thaon Di Revel and *serg.* Domenico Cacciatori with gunner Buratti. They fought off a Fokker three times which was finally driven down toward Ronziano, the badly damaged Ca 1157 landed at Chiasiellis airfield.

On 29 September the unit attacked Trieste again with Ca 1244 of Scavini, Montegani and Olivieri and Ca 1219 of Mongardini, Baccili and *ten.* Angelo D'Angelantonio, observer. In this period the unit partially renewed personnel and equipment, giving away its Ca 1142 and 1229 and receiving from Taliedo Ca 1244 and 1200, the latter a 350 hp model. Mongardini was shifted to the 4ª Squadriglia while *sottotenenti* Domenico De Lorenzi and Enrico Bertolini arrived.

On 6 December 1916 *cap.* Rossi was assigned to the Comando d'Aeronautica and *cap. pilota* Giulio Sibilla Massiera became the new commander. During 1916 the 6ª Squadriglia flew 36 combat sorties. Silver Medals were assigned to *serg.* Minardi and Montegani, *s.ten.* Olivieri, *cap.* Rossi and *s.ten.* Scavini, Bronze to *serg.* De Rossi.

In the first three months of 1917 there was no activity, due to bad weather. On 10 April 1917 the 6ª Squadriglia was assigned to the newly formed XI Gruppo at Aviano. Bad weather precluded flying until May, when the 10th offensive of the Isonzo began. The unit then had on strength Capronis Ca 1244, 1229, 1231, 1250 and its force consisted

Left: Observer *ten.* Mario Olivieri with Ca 300 hp 1142 of the 6ª Squadriglia.

of *cap.* Sibilla Massiera, pilots *s.ten.* Thaon di Revel, Scavini and Enrico Bertolini, *asp.* Cesare Baccili, *serg.* Cacciatori, Montegani and Angelo Pacassoni; observers *cap.* Giulio Schiller, *tenenti* Mario Olivieri, Angelo D'Angelantonio, Ferruccio Pizzi; gunners Aldo Buratti, Ciastellardi, Luigi Ferippi, Paolo Mascagni, Bernado Picco, Alfredo Porta, Paolo Raccagni and Nazzareno Salvi.

On 2 May all four bombers flew a night mission against Opcina. A thick fog hampered their return and Ca 1229 got lost and made an emergency landing near Fontaniva, where it was wrecked with no harm to its crew. The new Capronis 450 hp arrived only in May.

In June 1917 the Squadriglia operated on the Trentino front, supporting the bloody failed attack on Mount Ortigara. On 19 June the unit bombed Albergo Ghertele together with other squadrons. On 28 July its Ca 450 hp 2361, 2393, and 2398, just recently delivered to the Squadriglia, attacked Idria.

On 2 August 1917 four bombers took off at midnight heading for Pola to bomb the Arsenal, Scoglio Ulivi and the fleet at mooring. *Ten.* Ferruccio Pizzi and *Guardiamarina* Umberto Calvello participated in this mission as observers. The latter, belonging to the 251ª Squadriglia of the Regia Marina at Venice, flew on Ca 2393 with pilots Thaon di Revel and Pacassoni. This mission was repeated the next night. In the last two weeks of August the unit participated in the massive Italian air effort, hitting Cerovlje, Tennizka and the Hermada Hill. On 19 August enemy artillery badly

Above: The photo of Caproni 4062 of the 6ª Squadriglia shot down on 26 October 1917 was circulated for propaganda purposes.

damaged the Caproni of *cap.* Sibilla. On 23 August Caproni 2399 burned in its hangar on account of an accident. At the end of the summer Baccili and Pacassoni were assigned to the Distaccamento A.R. preparing the raid against Cattaro. On Sept.1 Caproni 4137 was tested on a long-range flight to Turin and back, which was cut short due to engine problems.

After the enemy breakthrough, on 25 October the 6ª Squadriglia sent its Ca 4076 and 4062 in a night attack against German forces at S. Lucia di Tolmino. The same aircraft repeated their attack the next day: Ca 4062 took off at 2.30 p.m. with Baccili, now *aspirante ufficiale*, *cpr.* Pietro Cattai, gunners Ferippi and Eugenio Violi, but failed to return. The crew of another bomber reported seeing it going down. The Caproni was probably victim of the Albatros of *Offizierstellvertreter* Schmidt of Jasta 39: German fighters claimed four Capronis that day. There were no survivors.

On 31 October 1917, the unit retreated to the airfield at Padua with Caproni 4076, 2393, 4137, and 2323, leaving behind unserviceable 2335. On 1 November the 6ª Squadriglia was reinforced with airplanes (serials 4170 and 4071) and crews from the Comando XI Gruppo, to which 6ª belonged, including *maggiori* Egidio Carta and Armando Armani, who often led the squadron into action. In this period therefore the 6ª Squadriglia had Ca 4170, 4071, 4076, 2393, 4137 and 2323. Its pilots were *cap.* Sibilla Massera, *tenenti* Bertolini, Montegani, Leandro Negrini, Luigi Ridolfi, *s.ten.* Pacassoni, *serg.* Cacciatori, *caporali* Ubaldo Puccione, Carlo Bonalumi and Domenico Guarino, its observers were *cap.* Schiller, *tenenti* Pizzi, Francesco Cutry and Liborio Petracalvina, its gunners Buratti, Ettore Candiani, Tarcisio Cantarutti, Giuseppe Dal Lago and Ricciotti Lupani. Luigi Ridolfi was an excellent pilot, formerly instructor at Malpensa, under the protection of *magg.* Armani, chief of the Raggruppamento Squadriglie da Bombardamento, he passed through five different squadrons in the first five months of 1918.

Despite bad weather they attacked, by day and night, military targets at S. Giorgio Nogaro, Primolano and Ghertele. On 21 November Ca 4170 was damaged in a forced landing. On 8 December Capronis 4071, 4137, 4076, 2393 and 2332 took off from Padua and joined the Capronis from San Pelagio flying to bomb barracks and troops in the valley of Campomulo, then they went down to strafe enemy columns. On 13 December the 6ª Squadriglia moved again, to Ghedi, with its Ca 4076, 2393 and 2332. Ca 4137 stayed at Padua; it was destroyed on that field by enemy bombs on 3 January 1918. On 14 December *magg.* Armani and the personnel from the Comando di Gruppo left, taking with them Caproni 4071. The Squadriglia was temporarily deactivated. During 1917 it had flown 138 combat sorties. 13 Silver and 5 Bronze Medals were assigned to its personnel in 1917.

On 29 January 1918 the 6ª Squadriglia moved from Ghedi to Verona and on 5 February it got Ca 2376 arriving

Above: A Caproni at Ghedi airfield in the spring of 1918, with roundels and camouflage over the top wing.

Above: Caproni 4076 of Marelli and Soliani after its dramatic landing on the Adamello glacier, 25 May 1918. (Archive Quelli del 72)

from Milan. Operations were resumed on 6/7 February with a night bombing of the railroad station at Mattarello by Ca 2376 and 4076 with *sottotenenti* Manlio Borri and Ariboldo Soliani as crew chiefs. Ca 4076 got lost and made an emergency landing in the parade grounds of Ferrara, suffering no damages and returning to base the next day. In the following days the bombers attacked bridges near Bolzano and the airfield of La Comina. Four Capronis bombed the station of Mattarello three times, on 28 February, 25 and 29 March. In March Sibilla Massiera left, *ten.* Cutry was acting commander for a few days, then *ten.* Omero Cavallarin replaced him. At the end of March, the 6ª Squadriglia had Ca 11502, 11503, 4076 and 2376. On 25 March they attacked again at night the station of Mattarello, using their Ca 11502, 4076, 11503, and 11507, the last one borrowed from the 4ª Squadriglia as 2376 was unserviceable.

In this period, it had pilots *tenenti* Borri, Cavallarin, Marelli, Soliani, *sergenti* Bonalumi, Cacciatori, Guarino, Pacassoni and Puccioni, observers *tenenti* Cassinelli, Cutry, Fausto Da Passano, Petracalvina, *s.ten.* Angelo Pocetta. Its gunners were Alberto De Rossi, Andrea Lovera, Pietro Pezzoni and Giuseppe Vittone.

On 4 May 1918 the unit bombed powerplants at Fies. On 20 May the unit moved from Verona to nearby Cà degli Oppi, and remained there until 18 June. In that period "Nel Cielo", the excellent Italian aviation magazine, published on its cover the photograph of *ten.* Ridolfi, the winner of the competition *"Bombardieri del Cielo"*, (sky bombers) a prize awarded according to the number of sorties, flight hours, difficulties and tonnage dropped in the period from 1

Above: The new Caproni Ca.5 600 hp at Tombetta airfield, Verona, in October 1918. (Archive Caliaro)

Right: Lieuts. James Bahl and DeWitt Coleman. (Friends of the Foggiani)

November 1917 to 28 February 1918. Ridolfi had a credit of 24 combat sorties for a total of 2,855 minutes, which earned him a prize of 2,000 Lire. Ridolfi died with Resnati on 2 August 1919 flying a civilian Ca 600 loaded with fifteen journalists.

On 25 May a bombing mission over the Tonale Pass failed because of an extremely strong wind and Ca 4076 of *s.ten.* Soliani and *ten.* Sandro Marelli, observer *ten.* Guglielmo Cassinelli, was tossed around over the Adamello glacier, spinning and falling. Marelli and Soliani managed to land the Caproni in a small iced plain, where it stuck its wheels in the snow and turned over. Alpine troops rescued the unharmed airmen, and the pilots were officially praised for a remarkable landing at an altitude of 3,150 meters.

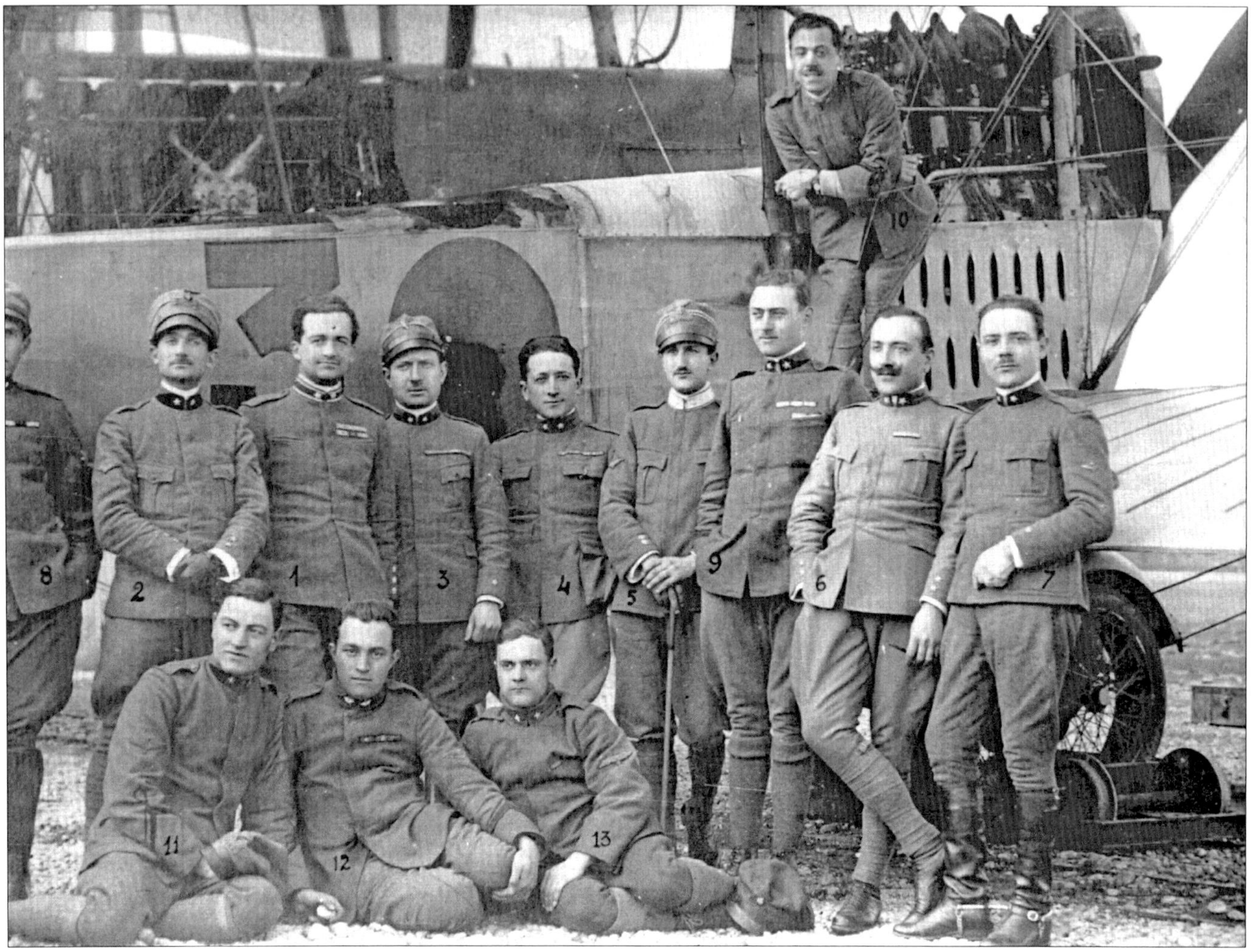

Above: The airmen of the 6ª Squadriglia with a Caproni Ca.5.

On 8 June four bombers attacked Pergine airfield and the station at Conegliano on 24 June.

At this time the 6ª Squadriglia had Ca 11502, 11503, 2338, and 2377. On 11 July 1918, tasked with a leaflet dropping mission, Caproni 2338 crashed on take-off, killing all its crew, *ten.* Soliani, *ten.* Marelli, the observer *s.ten.* Domenico Martino and gunner *sold.* Pietro Pezzoni. On 17 July the unit received its first American personnel coming from the flying school at Foggia: Lieutenants James L. Bahl Jr., Fenafly DeWitt Jr. Coleman, Arthur D. Farquhar, William B. Shelton, Norman Sweetser and Frederick K. Weyerhauser. In one of their first missions an explosive shell wounded Lt. Farquar, second pilot on Caproni 11502, in the head.

On 22 July *ten.* Cavallarin passed command to *cap.* Raffaele Tarantini. On 13 August personnel of this unit went to Taliedo to pick up the new 600 hp Caproni Ca.5. On 17 September Caproni 11503 and 2377 bombed at night the airfields of Aviano and Pergine, carrying Lieuts. Weyerhauser, observer, and Coleman and Bahl, pilots. On the 25th all three serviceable bombers of the unit, Ca 2377, 11503 and 4058, dropped leaflets in the area of Mezzolombardo.

The new Caproni 5 arrived on 4 October from Taliedo, they had serials 11647, 11664 and 11671, that got the insignia of the unit, a red sun, and two days later the Caproni Ca.3 450 hp 4086, 2377 and 11503 were turned over to the 4ª Squadriglia. The 6ª was thus the first bomber unit wholly equipped with this controversial bomber, with the plan of fielding eight of them, with 16 pilots. By the end of the month the 6ª Squadriglia, that moved to Padua on 22 October, had Caproni 5 600 hp serials 11647, 11664, 11671, 11669, 12214, 11659, 11676, and 11660. With commander Tarantini were now the pilots Lieuts. Bahl, Coleman, Sheldon, *ten.* Bonalumi, Borri, Cavallarin, Mario Giuliani Gusman and Vincenzo Machi, *s.ten.* Pacassoni, *serg.* Guarino and Domenico Greselin; observers Lt. Weyerhauser, *ten.* Petracalvina, Vincenzo Cutello, Francesco Cutry, *s.ten.* Orazio Aiello, Domenico Girardi and Cesare Morresi; gunners Cantarutti, Vittone, Alcibiade Borghi, Capperino Campanini and Mario Giampaoli.

On 27 October six bombers took off at 10.40 a.m. from Tombetta to bomb Vittorio Veneto during the Italian breakthrough. One of them didn't return, Ca 11669 crewed by the American pilots Lieuts. Coleman and Bahl with the Italians *ten.* Cutello and *serg.* Cantarutti. It was later learned that five Austrian fighters shot down the Italian bomber near Pergine, with the death of the whole crew. Among the Austrian fighters were *Oblt.* Roman Schmidt and Emmerich von Horvath. Lt. DeWitt Coleman Jr. was posthumously awarded the *Medaglia d'Oro al Valor Militare*, Italy's highest decoration.

On 1 November Ca 11664 of pilots *ten.* Borri and Lt. Shelton was forced to fly at low level by clouds and ground fire badly damaged it, it still completed its mission of bombing bridges at Latisana and with a single engine running it made an emergency landing still under enemy fire, with no harm for its crew. The final action of the war was a bombing sortie against the railroad near Palmanova Ca.5 11647 and 11660 on 2 November 1918. During 1918 the 6ª Squadriglia had performed 147 combat sorties with 25 of them with the Caproni Ca.5, bringing the total for the whole war at 319, in 144 war missions, dropping 75,915 kgs. of bombs and 6,750 kgs. of leaflets. One Gold Medal, twenty Silver and four Bronze medals were assigned to its airmen in 1918. It was disbanded in 1919.

Casualties and Combat Losses of the 6ª Squadriglia

Date	Place	Aircraft	Crew	Cause
26.10.1917	S. Lucia	Ca 4062	*Asp.* Baccili KIA, *cpr.* Cattai KIA, *sold.* Ferippi KIA, *sold.* Violi KIA	Fighters
11.7.1918	Tombetta	Ca 2338	*Ten.* Marelli KIC, *ten.* Soliani KIC, *s.ten.* Martino KIC, *sold.* Pezzoni KIC	Accident
27.10.1918	Vittorio	Ca 11669	Lt. Bahl KIA, Lt. Coleman KIA, *ten.* Cutello KIA, *serg.* Cantarutti KIA	Fighters

Above: A Caproni Ca.5 of the 6ª Squadriglia

7ª Squadriglia

Above: The emblem of the 7ª Squadriglia.

This Squadriglia was formed at Aviano airfield on 12 January 1916 as 7ª Squadriglia da offesa, with commander *cap.* Domenico Mondelli and as part of the Gruppo Comando Battaglione Aviatori assigned to the Supreme Command. Its components were pilots *cap.* Piero Gregorini, *cpr.* Gaudenzio Signorelli, *sold.* Amerigo Contini and it was equipped with Ca 300 serials 487, 715 and 719.

After a standing patrol above Udine, the 7ª Squadriglia had its operational debut participating in the bombing of Ljubljana on 18 February with its Ca 487 with *cap.* Gregorini and Signorelli as pilots. On 31 March the same Ca 487 with Mondelli and Signorelli, the observer *ten.* Umberto Fioretti and gunner Oreste Negri flew a reconnaissance mission north of Adelsberg together with aircraft of the 6ª Squadriglia. On 5 April Ca 715 with pilots *ten.* Adriano Bacula, (future test pilot of SIAI), and *serg.* Menenio Bosio and gunners Riccardo Picco and Ettore D'Arduin bombed Adelsberg. On 20 April Ca 719 with the same pilots and as observer *Tenente di Vascello* Luigi Bologna, bombed the seaplane yards at Trieste, while Ca 715, with *cap.* Mondelli, *serg.* Signorelli and the Frenchman EV André Vaugeois, of the seaplane squadron in Venice as pathfinder, crashed near Aquileia for engine failure, Vaugeois being injured.

Meanwhile, on 16 April, the unit was renamed from 7ª Squadriglia da Offesa to 7ª Squadriglia Caproni, while the Gruppo Comando Battaglione Aviatori became IV Gruppo.

On 16 May Ca 719 of Bacula and Bosio, with gunners Negri and D'Arduin, took off from Aviano in the early morning to bomb barracks and railroads at Castagnevizza (Kostanjevica). As it was dropping its bombs, a Fokker and two biplanes attacked it, in the ensuing gunfight one of the latter was seen diving steeply down. The bomber returned to base. In June and July 1916, the unit operated on the Trentino front against the enemy offensive. It attacked barracks in the valleys of Campomulo and Nos, Albergo del Ghertele, Osteria del Termine and the airfield of Cirè.

On 1 August 1916 two bombers of the 7ª Squadriglia flew in the raid against the torpedo factory in Fiume. Ca 1161 was forced back by bad weather, but Ca 719 managed to reach the city and to place two bombs on its target. On 7 September *cap.* Mondelli was reassigned to the Comando di Aeronautica and *cap.* Vittorio De Muro replaced him. At this time the squadron had Caproni 719, 1161, 1236 and 1255. Its pilots were *s.tenenti* Francesco Broili, Mario Contini and Riccardo Ravioli, *sergenti* Bosio and Signorelli, *cpr.* Alberto Barducci and *sold.* Armando Manicardi with the observers *tenenti* Baviero Baviera and Mario Condulmari, gunners D'Arduin, Negri, Achille Vincenzo De Lama and Beniamino Lattanzio.

On 13 September 1916 all the four aircraft of the 7ª Squadriglia took part in the mass bombing of Trieste arsenal, dropping 26 162-mm bombs. On 14 October the unit transferred its Ca 350 hp 1255 to the 2ª Squadriglia and received Ca 300 hp 1160. On 31 October Caproni 1236 and 1161 raided Dottogliano, near Trieste. On their return, at an altitude of 3,000 meters, two Austrian fighters attacked Ca 1161 of Ravioli, Bosio and Condulmari above Nabresina. All three airmen were wounded, the engines failed, but the pilots managed an emergency landing in the marshes near Punta Sdobba. The Caproni was shot down by Gottfried Banfield, flying the seaplane fighter A.3, together with *Hpt.* Viktor Schunzel, commander of Flik 28.

In November 1916 the 7ª Squadriglia supported the 9th Battle of the Isonzo, another bloody and fruitless offensive. On 3 December Caproni 1236 and 719 were tasked again with bombing the station of Dottogliano. Over Comen two fighters attacked the returning Ca 719 of Broili, Barducci and D'Arduin, putting an engine out of action. The bomber went down and Broili managed an emergency landing near Cormons, without harm for the occupants. Winners of this combat were the Austrian ace Adolf Heyrowsky flying a Brandenburg C.1 with *Fähnrich* Stefan Wagner, in co-operation with another two-seater of *Feldwebel* Heinrich Mahner and the observer *Lt.* Alexander Tahy, who was slightly wounded by the bomber's defensive fire. *Serg.* Barducci was credited with an enemy aircraft driven down.

Above: Men of the 7ª Squadriglia in front of Caproni 719, which was shot down on 3 December 1916.

At the end of January 1917, the 7ª Squadriglia, still at Aviano assigned to the IV Gruppo, had commander *cap.* De Muro, pilots *ten.* Ravioli, *s.ten.* Quintilio De Medio, Cesare Dima and Antonio Magnacavallo, *serg.* Bosio, *cpr.* Barducci, *sold.* Manicardi; observers *ten.* Baviera, *cap.* Tito Giovambattista Benetti and Adamo Sturlese; gunners D'Arduin, De Lama, Lattanzio, Negri and Amedeo Faoro. Its equipment consisted of Caproni 300 hp 719, 1232 and 1236.

On 7 March 1917 the IV Gruppo tasked its 7ª Squadriglia to fly the first night bombing mission. Three Capronis of this unit attacked barracks and warehouses near Reifenberg, defying searchlights and a strong anti-aircraft fire, and earning the praise of the Comando di Gruppo. On 10 April the 7ª Squadriglia was reassigned to the XI Gruppo.

In May 1917 the unit supported the 10th offensive of the Isonzo hitting the Austrian lines and the airfield of Prosecco, north of Trieste. In June the 7ª Squadriglia flew in support of the futile and bloody attack on the Ortigara.

Its personnel then included cap. De Muro, *ten.* Riccardo Ravioli, *s.ten.* Magnacavallo, De Medio, Dima, *serg.* Barducci and Menenio Bosio, observers *cap.* Tito Benetti, *ten.* Baviera, *s.ten.* Sturlese, gunners *cpr.* Oreste Negri, *soldati* Siliano Berti, D'Arduin, De Lama, Giuseppe Brugo, Lattanzio.

By 1 August 1917 the 7ª Squadriglia had Capronis 450 hp 2399, 2400, 2402 and 2403 and the same personnel as in January, less Magnacavallo and with *sold. pilota* Silvio Geri.

On 3 August three bombers attacked the arsenal of Pola by night, but on the way back Ca 2400 of *ten.* Ravioli, *serg.* Bosio, Benetti and D'Arduin got lost in the clouds and the fog, and its pilots opted for an emergency landing in the bed of the Adige River, near Legnago. Their airplane hit a telegraph cable and flipped over in the water, but the four airmen managed to get out. During the massive operations of the battle of the Bainsizza, the 7ª Squadriglia operated against Austrian emplacements at Grahovo, Ternizza and over Hermada Hill even twice in the same day.

In the nights of 3 and 4 September this unit attacked the arsenal of Pola together with Capronis of other Squadriglie. On 5 September Caproni 2403 with *ten.* Ravioli, *serg.* Bosio, *cap. oss.* Benetti and *cpr.* D'Arduin failed a take-off for a sortie against Sistiana and crashed at the end of the airfield, killing all its crew of veterans. On 20 September gunner Amedeo Faoro and pilot *sold.* Manicardi, the observer *ten.* Raffaele Vallarino and *ten.* Giovanni Strata were assigned to the *ad-hoc* 15ª Squadriglia Bis part of the Distaccamento

Left: Another view of Caproni 719; over the serial number there is the ribbon of the medal *"Distintivo speciale per i militari ed assimilati che sopportano le fatiche dell'attuale Guerra"* (Special badge for soldiers and related personnel who endure the hardships of the current war).

A.R. for the planned sortie against Cattaro. This unit attacked Pola twice again, on 27 September and 2 October.

After the enemy breakthrough the 7ª Squadriglia on 25 October bombed S. Lucia di Tolmino, and in the following days it attacked the bridges on the Torre River and Codroipo. In that period *cap.* De Muro was shifted to the Deposito Aeronautico, the observer *cap.* Luigi Leoncini replaced him as commander. The Squadriglia first moved to Marcon on 31 October, then on 7 November to Ghedi, saving all its aircraft: 2399, 2402, 4042 and 4056.

Still with *cap.* Leoncini as commander, the 7ª Squadriglia was assigned to the XIV Gruppo. This unit now had on strength Caproni 2399, 2402, 4042 and 4056. Its pilots now were *ten.* Dima, *s.ten.* De Medio, *serg.* Barducci and Manicardi and *cpr.* Lorenzo Sobrero; its observers *tenenti* Baviera, Sturlese, Vallarino and Piero Rosalba; its gunners: Brugo, Faoro, Lattanzio and Dante Barbuti.

On 4 January 1918 the unit moved again, and settled at Padua airfield. On 25 January it was tasked to bomb Primolano, a staging point threatening the Italian positions on the Grappa Mountain. In the night Ca 2402 with *ten.* Giovanni Saccani, *cpr.* Sobrero, *ten.* Rosalba and gunner Barbuti and Ca 4056 with *ten.* Giovanni Cristiani, *serg.* Manicardi, *ten.* Sturlese and *cpr.* Faoro took off, but when they returned, they found heavy fog, and on landing Ca 2402 hit a bomb dump and crashed. Luckily there were no serious consequences and Sobrero and Rosalba were only slightly injured.

On 23 February 1918 the unit bombed La Comina airfield, now held by the enemy, and all through February and March it continued its actions against Primolano, Bolzano, Caldonazzo and Portobuffolè. In May and until early June it attacked the airfields at Feltre, S. Pietro in Campo and Godega di Sant'Urbano. In this period commander of the 7ª Squadriglia was *cap. pilota* Filippo Aliquò Mazzei; the unit used Caproni Ca.3 11497, 11498 and 11501. Its force included pilots *ten.* Saccani, *s.ten.* Euclide Cayre, *serg.* Manicardi and Giuseppe Bernardi, *cpr.* Giovanni Baldini and *sold.* Giovanni Ferrero; the observers: *tenenti* Leoncini, Sturlese, Gino Carando and Arturo Cuscianna, *s.tenenti* Pietro Amadessi and Marco Visconti; gunners: Barbuti, Faoro, Lattanzio, Vittorio De Villoni and Corrado Milani.

During the Austrian offensive of mid-June, the squadron was active particularly against the bridges and causeways across the Piave River, bombing Falzè and the meadow of Zenson. On 23 June ground fire damaged Ca 11501, *cpr.* Faoro managed to repair the carburetor of an engine. In the afternoon of 24 June Caproni 11497 and 11498 took off from Padua to attack the station at Pramaggiore. The first one, with pilots *s.ten.* Edgardo Farina and *serg.* Bernardi with the observer Cuscianna and gunner De Villoni, claimed a victory over an attacking fighter. The bomber returned to its airfield with 162 bullet holes.

On 22 July the 7ª Squadriglia was deployed at the airfield of Arquà Petrarca, where it remained until the end of the war. During the summer this unit received the American pilots Lieuts. Arthur M. Beach, Gilbert P. Bogert, William G. Cochran, Alexander M. Craig, Edmund Dorman and Harry S. Finkenstaedt. In the night of 23 August Capronis 4042, 11497 and 11501 bombed the airship hangar at Casarsa della Delizia. In the last days of October during the final offensive the 7ª Squadriglia bombed barracks and troops at Fonzaso and Sacile, then it attacked bridges and roads, as the enemy was being routed. Its final mission was an attack against the bridge over the Tagliamento at Latisana on 1 November 1918.

At the end of the war the 7ª Squadriglia, with commander Aliquò Mazzei, had pilots Lieuts. Beach, Cochran, Dorman, Finkenstaedt; *ten.* Saccani and Cayre; *s.ten.* Farina and Alfredo Alderighi; *serg.* Baldini, Bernardi and Giacomo

Above: Before being assigned to the 7ª Squadriglia, Caproni 1160 was tested with a 25-mm cannon. How this gun could be operated in flight is not clear.

Rossi; observers *s.ten.* Amadessi, *ten.* Carando, Leoncini, Sturlese and Vallarino; gunners Lattanzio, Arturo Cabrini, Francesco Gagliardi, Giuseppe Zappoli, Francesco Zavattaro.

This squadron, that during the war had flown 121 combat sorties, its men earning 13 Silver Medals and three Bronze Medals, was disbanded on 28 February 1919.

Casualties and Combat Losses of the 7ª Squadriglia

Date	Place	Aircraft	Crew	Cause
20.4.1916	Aquileia	Ca 715	*Cap.* Mondelli UNH, *serg.* Signorelli UNH, EV Vaugeois WIC	Accident
31.10.1916	Punta Sdobba	Ca 1161	*Ten.* Ravioli UNH, *serg.* Bosio UNH, *ten.* Condulmari WIA	Fighters
3.12.1916	Oppacchiasella	Ca 719	*Ten.* Broili UNH, *cpr.* Barducci UNH, *cpr.* D'Arduin UNH	Fighters
5.9.1917	Aviano	Ca 2403	*Ten.* Ravioli KIC, *serg.* Bosio KIC, *cap.* Benetti KIC, *cpr.* D'Arduin KIC	Accident

Above: Caproni 1232, in an unusual dark finish, at Aviano. in the background two other Capronis of the 7ª Squadriglia.

Above: A Caproni Ca.3 of the 7ª Squadriglia at Aviano.

Above: Caproni Ca.3 of the 7ª Squadriglia, probably at Arquà Petrarca.

Air-to-air photo of a Caproni of the 7ª Squadriglia, in dark finish; the blue stars appear also on top of the fuselages, there are no roundels on the wings. (Archive Fiorenzo Longhi)

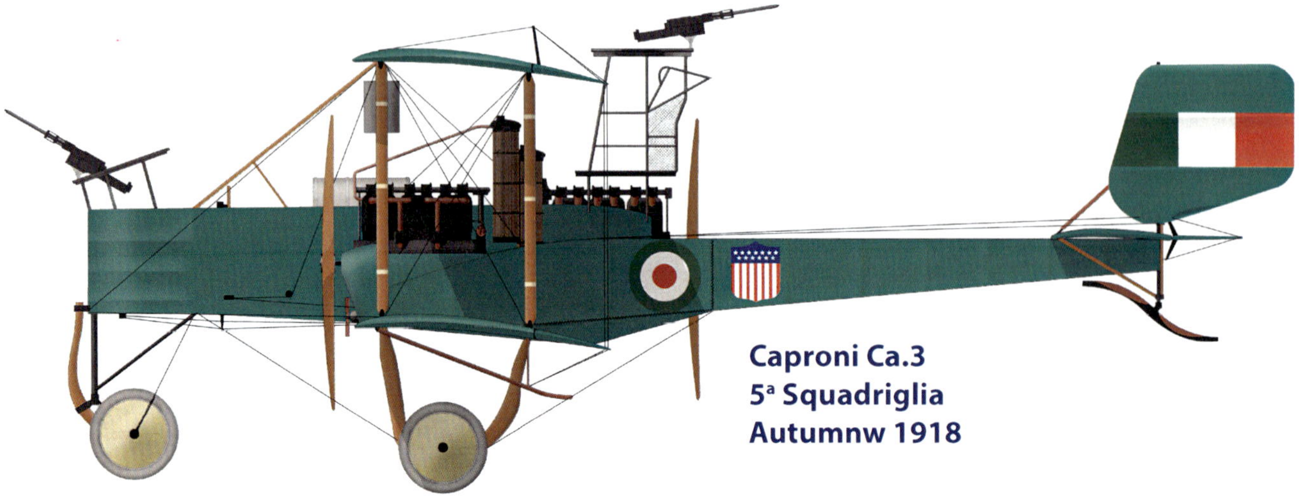

Caproni Ca.3
5ª Squadriglia
Autumnw 1918

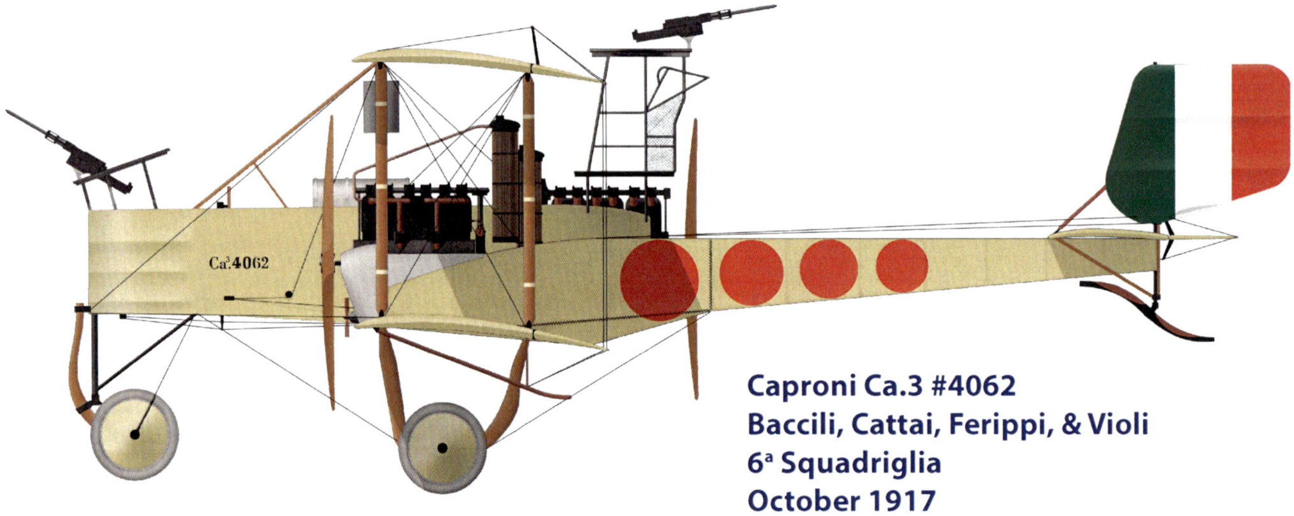

Caproni Ca.3 #4062
Baccili, Cattai, Ferippi, & Violi
6ª Squadriglia
October 1917

Caproni Ca.5
6ª Squadriglia
October 1918

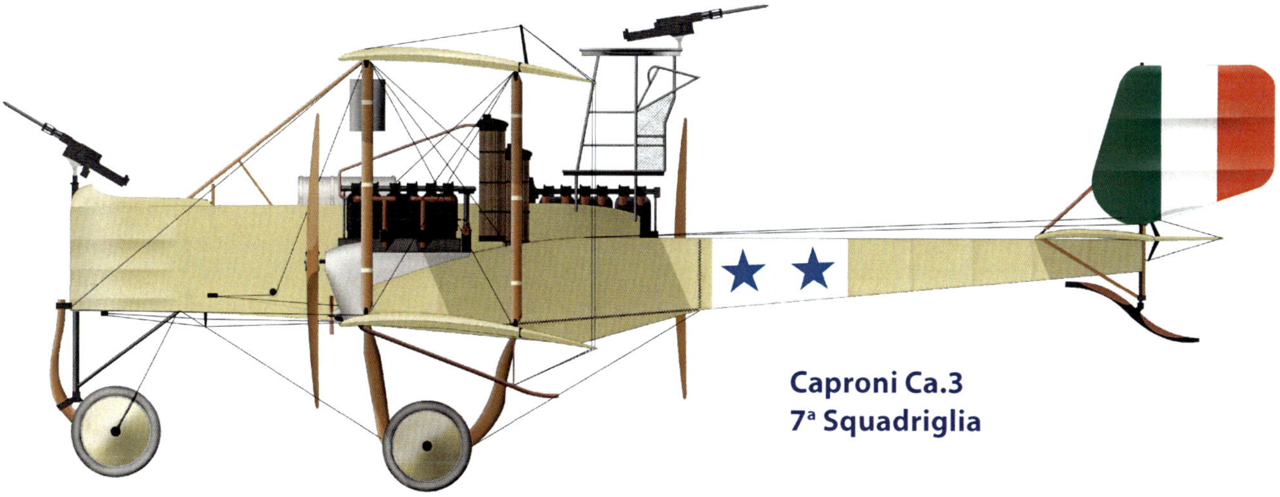

Caproni Ca.3
7ª Squadriglia

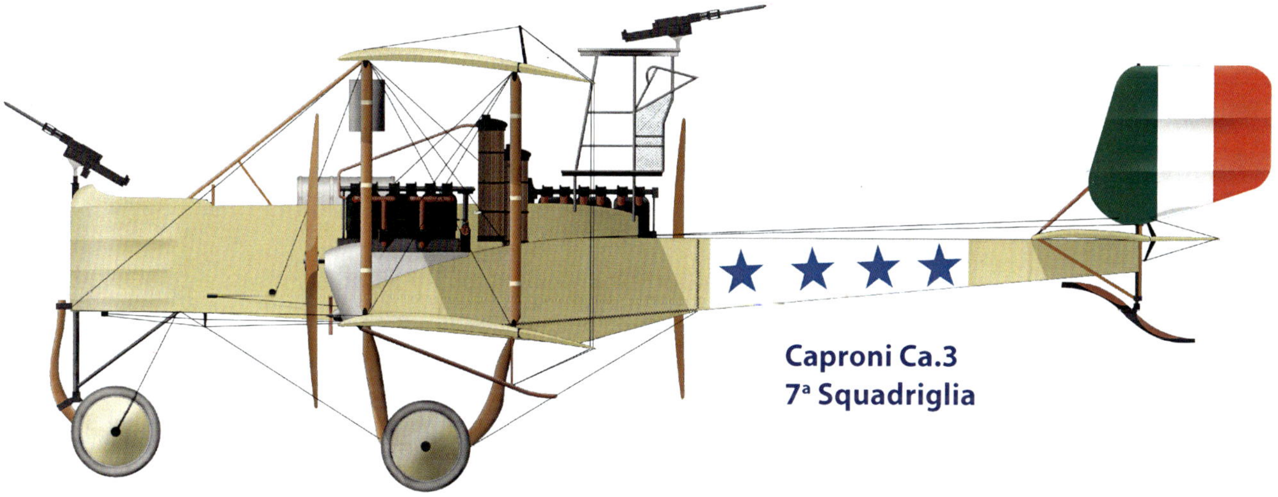

Caproni Ca.3
7ª Squadriglia

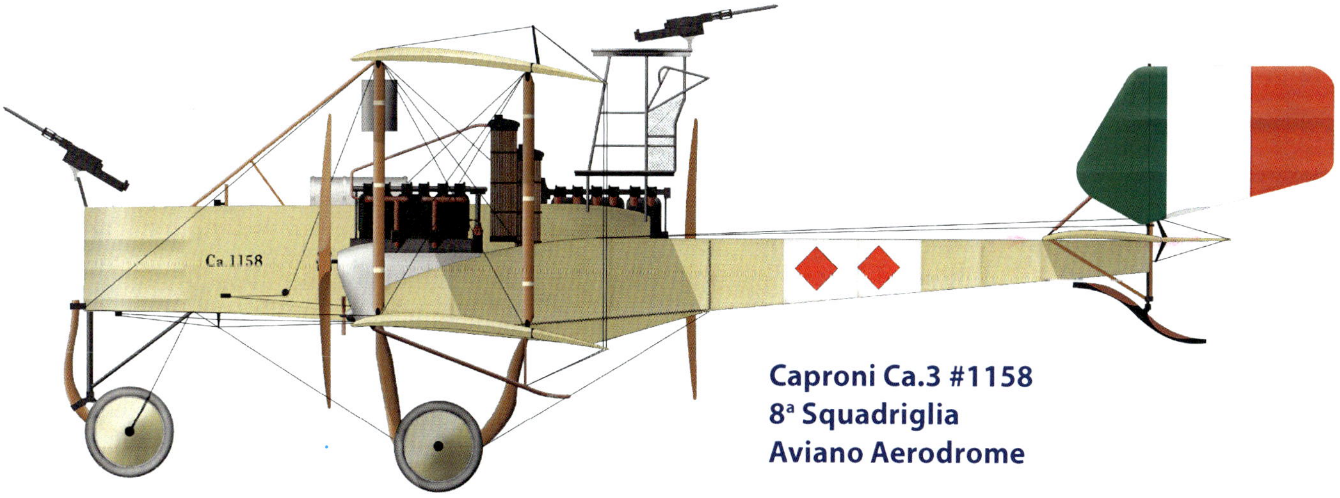

Caproni Ca.3 #1158
8ª Squadriglia
Aviano Aerodrome

8ª Squadriglia

Above: The first "ace of clubs" of the 8ª Squadriglia, Ca.1 1151.

The 8ª Squadriglia was formed on 25 March 1916 at La Comina airfield. It reported to the Comando Battaglione Aviatori (IV Gruppo since 15 April) and its commander was *cap. pilota* Luigi Govi, with the pilots *s.ten.* Mario Martini, *s.ten.* Giuseppe Romelli, *cpr.* Giacomo D'Odorico, *soldati* Alessandro Borello and Gino Lisa. The Squadriglia pilots ferried from Taliedo and Vizzola its first Capronis on 26 April: 1151, 1158 and 1218.

They were used for intensive training, and then on 16 May 1916 there was the first combat mission, two airplanes taking off to bomb Ovcia Draga, but the first one with *cap.* Govi and Borello was forced back with engines trouble and the second one was called back. Later on, with pilots Martini and Lisa, it flew a standing patrol over the Isonzo River lasting two hours. *Cap.* Marco Aurelio Barbarisi, observer, arrived from the 42ª Squadriglia.

The 8ª Squadriglia was much active during the combats on the Asiago front, attacking barracks and troops at Campomulo, Monte Erio and Campolongo with bombs and steel darts. On 25 May two Capronis dropped incendiary bombs in the area of Lake Caldonazzo with the purpose of setting fire to the woods. On 30 May it flew its first mission with three bombers, to Rotzo and Roana. On 20 June the Austrian airfield at Cirè, near Pergine, was bombed; Ca 1218 of *cap.* Govi and Borello with gunner Massimo Bianchetti and 1151 of Romelli, D'Odorico and gunner cpr. Stefano Sciaccaluga faced an attack of five Austrian airplanes and fought them off, one of the attackers was seen dropping out of the combat. The Italian bombers returned safely, full of bullet holes. No combat flights took place from 21 June to 12 July, when another mission was flown, and then until 31 July.

On 1 August 1916 the bombers of the IV Gruppo attacked the Whitehead torpedo factory at Fiume: the night before the airships M 1 and M 6 had tried to reach the same target but had to give up on account of a strong wind that also delayed departure of the bombers. The Capronis left in four groups, and the four planes of the 8ª Squadriglia, including Ca 1230, formed the first group with *cap.* Govi as leader, together with the Capronis of the 1ª Squadriglia. D'Odorico, second pilot on Ca 1151 of Romelli, was commended in the Daily Bulletin of the Comando Battaglione Aviatori because he went out on the wing to change a sparkplug of the left engine in flight. On 15 August two bombers of the squadron attacked the station

Above: Ca.1 1158 at Aviano.

of Dornberg. At the end of August, the 8ª Squadriglia had Capronis Ca 1151, 1158, 1218, 1230 and 1256. Its personnel were pilots *cap.* Govi, *cap.* Vittorio De Muro, *s.ten.* Martini and Romelli, *serg.* Borello, D'Odorico, Lisa and Giovanni Tasso; observer *cap.* Barbarisi and gunners Bianchetti, Giuseppe Giovannini and Alessandro Zamengo. The unit was heavily engaged supporting the 6th Battle of the Isonzo hitting Prvacina, Dornberg and Opicina.

On 13 September 1916 the seaplane yards at Trieste were targeted as a reprisal for Austrian attacks on Venice. One airplane failed to return, Caproni 1256 of *s.ten.* Romelli, *serg.* D'Odorico and *sold.* Evaristo Zanetti that fell in flames, possibly victims of an engine failure. On 17 September, Ca 1158 of *s.ten.* Martini and *cpr.* Lisa bombed Dottogliano. There were no more combat missions all through October, due to constant bad weather. On 31 October a long-delayed bombing of Dottogliano took place, with Ca 1151 and 1193. Ca 1151 had a new crew, *tenenti* Luigi Gori and Maurizio Pagliano, who later became the most famous Caproni pilots, with *cap.* Barbarisi.

Until the end of the year, the unit supported the useless offensives of the 2ª and 3ª Armata along the Isonzo. The Capronis of the 8ª Squadriglia were marked with the suits of cards. The one of *cap.* Govi was marked with an ace of diamonds.

On 1 November there was another mass attack on Ovcia Draga and Cernizza, the right engine of Ca 1193 of Govi, Borello and Bianchetti caught fire, the damaged bomber made an emergency landing at Campoformido airfield. Weather precluded further attacks, but on 15 November Martini, Lisa, Barbarisi and gunner *sold.* Zamengo took off on Ca 1158 to intercept Austrian airplanes, they didn't find them, the central engine failed, parts of the propeller struck the airplane, but they managed to return.

In early December 1916 the 8ª Squadriglia, still at La Comina, had pilots *cap.* Govi, *tenenti* Martini, Gori, Pagliano, *s.ten.* Amerigo Contini, *serg.* Borello, Lisa, and *cpr.* Alfredo Fiscali; the observer *cap.* Barbarisi and gunners Bianchetti, Zamengo and Silvio Cossu.

Winter forced inactivity until April 1917. The first mission was on 13 April, Ca 2609 of Pagliano, Gori, Barbarisi and Zamengo and Ca 1230 of Contini, Fiscali, observer *ten.* Giuseppe De Vizzi and gunner Abele Galbiati bombed Vogersko and Probacina. On 20 April, tasked for a mission to Chiapovano, Ca 1193 had to return to base and Ca 1230 had to force land at Borgnano, Ca 2609 of Pagliano and Gori completed the mission. On 25 April *cap.* Govi was assigned to the Direzione Tecnica of the Comando Aeronautica and *cap.* Cesare Darby became the new commander. In the night of 2/3 May in action against Opicina 1193 had to turn back

Above: Caproni Ca.3 2609 with pilots Gori, left, and Pagliano, right. The flags show the halberd, symbol of Trieste, and an Italian flag with St. George and the dragon, a gift of patriotic ladies.

again, Ca 1230 of commander Darby and *asp.* Alessandro Borello bombed Prosecco airfield, Ca 2609 of Pagliano and Gori bombed Opicina.

On 10 May 1917, Caproni 1230, 1193 and 2609 took off for a night sortie against barracks and the station at San Daniele. South of Duino, Pagliano, Gori and Barbarisi, flying Ca 2609, felt sick forming the opinion that they had met an anti-aircraft barrage of poison gas. They decided to bomb Pola as a reprisal and dropped ten bombs on the Austrian naval base, and the remaining four on San Daniele on the way back. In their report they wrote: "*For reasons of chivalry the crew chose to turn the lights of the plane on for a short time after dropping the bombs*". Their action, the first Italian bombing of Pola, was mentioned in the war Bulletin, but *magg.* La Polla, commander of the Raggruppamento, jailed Barbarisi for eight days on account of his disregard of orders.

On 14 May barracks east of Gorizia were bombed, Ca 2612 of Darby and Contini and Ca 1183 had to abort for problems with the engines, Ca 2609 alone completed the mission, Pagliano and Gori having Pratesi as observer, and then remained half an hour strafing enemy troops to give a moral boost to the Italian infantry engaged in battle. On 18 May Darby flew Ca 2609 with *magg.* Falchi, the Group commander, with Ca 1193 of Martini and *serg.* Luigi Barberi, enjoying the escort of Nieuport fighters. Another three planes mission took off on 23 May, with Ca 1230 of Darby and 2609 of Pagliano and Gori turning back for engine troubles, and only 1193 of Borello and Fiscali completed the task and faced the attack of two Austrian fighters. At the end of May, Ca 1193, 2609 and 2612 were flown back to the factory and the squadron received 2378, 2381, 2309 and 2311. More missions followed on 5 June, 15 June, with four planes taking off, 19 June, night of 25/26 June, a moonless might, Pagliano and Gori on 2311 bombed Prosecco, testing the procedures for such flights.

8ª Squadriglia took part in the attack on Idria of 7 July with three planes in close formation. The mission of two Capronis to Chiapovano in the night of July 12 included the dropping of leaflets.

For a day action to S. Daniele del Carso on 23 July Darby flew with *ten.* Mario Muratorio on Ca 2309, but as the plane had troubles, they landed and took off with another Caproni after 2378 and 2381. During this raid, some bombs fell on Italian troops, killing 13 soldiers, causing the dismissal of *magg.* Falchi and *magg.* Mondelli, commanders of the IV and XI Gruppi.

The action against the power plant at Idria was repeated on 28 July: Darby and Contini on Ca 2309 took off and

Above: Ca.3 4609: when empty and parked, the Caproni rested on its tail skids.

landed five times, for the usual engine troubles, until they gave up. 2378 and 2381 completed the mission as Spads of 91ª Squadriglia kept Austrian fighters away.

In the night of 2/3 August 1917, 8ª Squadriglia took part in the night bombing of Pola with three airplanes. In Ca 2378 carrying the Ace of Spades insignia of Gori and Pagliano there was the poet Gabriele d'Annunzio, while Pratesi was in the place of the rear gunner. The action was repeated the next night, 3/4 August with four planes, Ca 2378 again had d'Annunzio, who at the time was 54 years old and one-eyed. On 7 August four Capronis bombed Chiapovano, on 9 August the target was again Pola. Ca 2378 again had the crew of d'Annunzio, Pagliano, Gori and Pratesi.

In the second half of the month, as the battle raged, the unit hit Chiapovano, Hermada and Aisovizza. Formations of four bombers attacked on 11, 14 and 18 August, three on 19 August twice, morning and afternoon, then again 4 in the morning of 20 August and four on 21 August, with Ca 2325 of *cap.* Darby carrying *magg.* La Polla. Three Capronis bombed on 22 and 23 August, two the next day and on 26 August, three on 28 August, all mission enjoyed the escort of fighters.

In this period the 8ª Squadriglia, at the command of *cap.* Darby, used Caproni 2325, 2309, Ace of Clubs, 2381 Ace of Hearts, 2378 and 2390. Its pilots were *tenenti.* Gori, Martini, Pagliano, Ugo Fabbri and Mario Trivellato, *s.ten.* Amerigo Contini and Ettore Trezzi, *aspiranti* Borello and Gino Lisa, *sergenti* Barberi, Silvio Brena, Luigi Gallareto and Mario Muratorio, *cpr.* Alfredo Fiscali; observers *ten.* Paolo Benigni and G. B. Pratesi, and gunners Cossu, Zamengo, Salvatore Baldassarre, Abele Galbiati, Serafino Bianchetti, Giovanni Gandolfi, Giorgio Sacco, Demetrio Tognetti and Battista Spreafico.

Pola was their target again in the night of 4 September, Ca 2309 of *ten.* Fabbri, *serg.* Muratorio, *sold.* Gallareto and Spreafico bombed its target, then on its way back an enemy seaplane attacked it from below. Its tracer bullets damaged the fuel lines; however, the Caproni managed to cross the sea but it ran out of fuel and crash-landed smashing its wings, but its crew was unharmed.

During the enemy breakthrough 8ª Squadriglia attacked bridges and roads at Pinzano and Motta di Livenza. On 25 October a Caproni failed to return from a mission to Grahovo in the morning, probably a victim of German fighters. Its crew of pilots *ten.* Antonio Rasi and *serg.* Silvio Breno, gunners Antonio Zuccolo and Ermenegildo Dell'Acqua was captured. That same day Ca 4050 made a forced landing at Campoformido due to an engine failure. A repair team reached it too late, as the bomber was torched on the 28th to avoid its capture. Caproni 2381 also was unserviceable and it was burned at La Comina airport before the retreat.

On 2 November the 8ª Squadriglia moved with its remaining Ca. 2326, 4063,4064 and 4085 to the field of San Pelagio, near Padua, from where it operated over the Grappa sector. On 7 November *Vzfw.* Hannemann and *Lt.* Von Kerssenbrock of Jasta 39 shot down over Meduna the Caproni of *ten.* Muratorio and *sold.* Galbiati with *ten.* Fabbri and *sold.* Spreafico. The Italian airmen were captured and the enemy promptly dropped a note reporting their fate.

In January 1918 the 8ª Squadriglia had on strength Capronis 4063, 4179, 4192, 4206 and 4224. Its pilots were *cap.* Darby, *ten.* Trivellato, *serg.* Gallareto, Guido Didonè and Luigi Panizzon, *cpr.* Gino Abati, observers *tenenti* Benigni, Felice Bianchi, Giuseppe De Leva and Antonio Landini; gunners: Cossu, Gandolfi, Tognetti, Giuseppe Daffonchio, Silvio Giolo and Carlo Lasi. In January this unit attacked Primolano, Cismon and Val Campomulo. In February it raided at night Primolano and la Comina.

Above: Line-up of the Capronis of the 8ª Squadriglia at Aviano.

In early March *cap.* Cesare Peraglie became the new commander. In May the 8ª Squadriglia attacked electric plants at Fies and Osteria Palù and the airfields at Feltre and S. Giustina. During the battles of mid-June, it attacked enemy troops at S. Donà and on the Montello Hill, on 18 June *ten.* Benigni was wounded in action by ground fire. On 18 June anti-aircraft fire wounded *ten.* Mario Simonini on Ca. 11505 but *t.col.* La Polla brought the plane down safely.

On 17 July 1918, the unit raided Pola again, for the first time in full daylight.

In August 1918 the 8ª Squadriglia was still at San Pelagio, equipped with Capronis 4041, 4206, 4192 and 11504. It had on force the American pilots Lieuts. Spencer L. Hart and Herman F. Krueger and *ten.* Mario Marasi, *serg.* Didonè, Gallareto and Panizzon, observers *ten.* Bianchi, Landini and *s.ten.* Aurelio Mosso; gunners Baldassarre, Daffonchio and Giolo.

On 1 September two aircraft took off to bomb the airfield of Aviano. Two Austrian fighters from Portobuffolè attacked Ca 4206 of Krueger and Gallareto, Landini and Baldassarre. The Caproni returned their fire and one of the fighters was driven down into the clouds, possibly hit, while the other one broke off the combat. The squadron was active in the preparation of the final offensive, hitting communication lines and the airfield at Mansuè. The final war mission of the 8ª Squadriglia was an attack in the afternoon of 30 October against the Sacile-Pordenone road.

At the end of the war the 8ª Squadriglia was at San Pelagio with Capronis serial 4041, 4151 and 4206. With commander Peraglie there were the pilots Lieuts. Hart and James H. Kennedy, *ten.* Marasi and Roberto Bonessa, *serg.* Didonè, Gallareto, Amerigo Zucchetti and Francesco Arbore, observers: *ten.* Bianchi, Landini, Mosso, *s.ten.* Giorgio Hack and *s.ten.* Renzo Hack; gunners Baldassarre, Daffonchio, Giolo, Luigi Avanti, Angelo Meroso and Gianbattista Terrani. This unit was not disbanded, and remained in active service after the war.

Left: From the left, *ten.* Pratesi, unknown, d'Annunzio, *ten.* Pagliano, and *cap.* Darby. The poet is wearing a black mourning band for the death of his friend Giovanni Randaccio.

Above: Ca.3 2378 at La Comina, in the center is d'Annunzio flanked by his pilots, Pagliano and Gori.

Below: A well-known photo that represents Italian aviation in the Great War: the Caproni, the most important airplane, Pagliano and Gori, two gallant pilots killed in action, and the pervasive influence of the poet d'Annunzio with his mottos: "*Nulla via invia*", no road is too hard, and "*Repetita iuvant*", it is useful to repeat. The flag with St. George is tied to the landing gear.

Above: In this line-up of the 8ª Squadriglia spades, diamonds, and clubs can be seen.

Casualties and Combat Losses of the 8ª Squadriglia

Date	Place	Aircraft	Crew	Cause
13.9.1916	Aquileia	Ca 1256	*S.ten.* Romelli KIA, *serg.* D'Odorico KIA, *sold.* Zanetti KIA	Accident or ground fire?
4.9.1917	Cordovado	Ca 2309	*Ten.* Fabbri UNH, *serg.* Muratorio UNH, *sold.* Gallareto UNH, *sold.* Spreafico UNH	Fighters
25.10.1917	Grahovo	Caproni	*Ten.* Rasi POW, *serg.* Breno POW, Zuccolo POW, Dell'Acqua POW	Fighters
7.11.1917	Meduna	Caproni	*Ten.*Muratorio POW, *sold.* Galbiati POW, *ten.* Fabbri POW, *sold.* Spreafico POW	Fighters

Caproni Ca.3 #2378
8ª Squadriglia

9ª Squadriglia

Above: Ca.1 300 hp 1212 was assigned to the 9ª Squadriglia.

The 9ª Squadriglia was activated on 22 May 1916 at Taliedo, immediately flying to Verona, assigned to the III Gruppo, for the 1ª Armata, operating in the north-western part of the front. Its commander was *cap.* Luigi Reggio with pilots *s.ten.* Giampiero Clerici, *s.ten.* Alberto Moresco, *m.llo* Felice Torelli, *serg.* Ignazio Sargian, *soldati* Giuseppe Bacchetta, Pietro Lista and Federico Molteni; observers *cap.* Guido Barutta, *cap.* Achille Sergardi, *tenenti* Alberto Livi, Luigi Sella and *s.ten.* Emanuele Parini; gunners: Gino Ghiretti, Secondo Bosisio and Biagio Paoletti. This unit was equipped with Caproni 300 hp serials 1212, 1213 and 1237.

It went into action as the enemy was on the offensive called "Strafexpedition", and on 24 May it bombed the railroad station at Calliano with two planes, pilots, Reggio and Tozzi, Moresco and Lista, had a very difficult flight in bad weather. One bomber attacked the same target on 31 May, one on 2 June and two in June 5, Ca 1213 of Reggio and Torelli with gunner Ghiretti and observer *ten.* Giuseppe Micheli faced the attack of three Austrian fighters above Rovereto and claimed one of them shot down. On 16 June two bombers were tasked to attack with bombs and steel darts enemy encampments between Camporovere and Asiago. The one of Moresco, Lista and Micheli suffered a total engine failure just across the lines and it managed to force land at Casoni Gualdo, where it hit a tree smashing its wings, while its crewmen were unhurt. On 29 June a Caproni was tasked with a mission against Pergine, and enemy guns targeted it effectively, wrecking the left engine and almost dislocating it from its mount, while the observer Micheli was wounded. The bomber returned with 37 holes, and was flown back to the Caproni factory for repairs. In June 9ª Squadriglia completed 28 bombing sorties on the northern front, 11 in July.

On 8 July three Austrian airplanes attacked the Caproni of Reggio and Torelli, but only damaged it.

Another misfortune befell Clerici and Lista with the observer *ten.* Santoro Fugalli and gunner Egidio Pellizzari on 4 August: an oil pipe broke off and fluttered in the wind, with its pressure gauge dangerously close to the central propeller, so Pellizzari took a walk on the wing and secured with tape the threatening pipe. The 9ª Squadriglia operated until the end of 1916 supporting Italian troops on the area of the Asiago Plateau, flying a total of 62 combat sorties. The Silver Medal was assigned to *cap.* Sergardi, *s.tenenti* Clerici and Moresco, the Bronze Medal to *cpr.* Lista.

The year 1917 began with a bad mission on 6 January, in which two bombers aborted and force landed, one at Brescia and the other one at Caprino Veronese, where it was written off. Bad weather much reduced the number of flights, with

Above: This unidentified Caproni, marked with collar tabs, shows ten daisies, the symbol of the 9ª Squadriglia, probably meaning he number of missions it flew.

just five sorties in the first five months of 1917.

In the mission of 26 March, the Caproni of *serg.* Ignazio Sargian and Lista had a duel with a Brandenburg, being hit repeatedly. Having now received Ca.3 450 hp 2364, 2372, 2377 and 2387, on 23 May 1917, the unit was deployed at Campoformido, to support the 10th offensive of the Isonzo. At that time the 9ª Squadriglia, still with commander *cap.* Reggio, included pilots *tenenti* Clerici, De Lorenzi, *s.tenenti* Lista, Molteni, Sargian, *serg.* Claudio Savini, *cpr.* Bogino; observers *tenenti* Fugalli, Mento and Umberto Malvano; gunners Ghiretti, Paoletti, Tullio Ambrosio, Giuseppe Ferraris, Pericle Micheletto, Arturo Palmiro, Domenico Rampini, Primo Rechetti and Agostino Sciaccaluga.

The 9ª Squadriglia moved to Padua on 1 June 1917, after having flown two missions from Campoformido, one of which against Brestovica, and was engaged again on the Trentino front. On 10 June three bombers of the 9ª Squadriglia attacked Campolungo and Mount Erio: Ca 2364 with pilots *s.ten.* Domenico De Lorenzi and *cpr.* Oreste Bogino, observer *ten.* Pasquale Mento and gunner Paoletti suffered an engine failure and made an emergency landing at Nove di Bassano, suffering some damage. At the end of July it received the order was to fly standing patrols over the railroad stations of Brescia and Salò to preclude Austrian reconnaissance of troop movements.

On 25 July Reggio passed his command to *cap.* Dante Jussa. On 7 August the Squadriglia moved back again to Campoformido to support the 11th Battle of the Isonzo, forming with the 5ª Squadriglia a Campoformido Detachment. On 16 August three of its planes bombed Comen; on 20 and 21 August eight sorties were flown against enemy targets. In the nights of 4 and 28 September five sorties were flown against Pola.

On 21 September *ten.* Clerici, *s.ten.* Molteni, *sold.* Primo Repetti and the observer *ten.* Giovambattista Lucci Chiarissi were assigned to the Distaccamento A.R. On 28 September the unit suffered its first and only casualties: Ca 2327, returning from a night bombing mission to Pola, crash landed and turned over, *cap.* Jussa was killed while *s.ten.* Sargian, *ten.* Fugalli and the gunner Ferraris were seriously injured.

On 29 September another night attack on Pola was the mission. Caproni 4083 of *ten.* De Lorenzi, Savini, Mento and Paoletti was targeted by searchlights and bursting shells, one of its tanks was pierced, then an Austrian fighter seaplane placed its tracing bullets into the bomber. Gunner Paoletti drove off the attacker. Somehow, they managed to return to base, but their bomber was so badly damaged that it was

Two pilots with a Caproni showing the daisy insignia of the 9ª Squadriglia. (Archive Moncalvo)

The insignia of the 9ª Squadriglia appears on this wrecked, unidentified Caproni.

Ca.3 4175 was another Caproni from the disbanded 201ª Squadriglia that was assigned to the 9ª Squadriglia.

Above: Caproni Ca.3 11495, loaded with bombs, is fitted with mudguards at the wheels. (Archive Roberto Mantiero)

struck off charge.

On 15 October 1917 the 9ª Squadriglia left the III Gruppo and was assigned to the IV Gruppo, still remaining based at Campoformido. During the operations after the enemy breakthrough the unit, having as temporary commanders first *ten.* Mento and then *cap.* Agostino Crosa di Vergani, bombed enemy forces at Grahovo and S. Lucia di Tolmino and dropped food to isolated Italian troops. On 2 November it moved from Campoformido to Arcade, with 4053, 4075 and 4083 having torched its Capronis 2392 and 4163, and then on 16 November it moved to Ghedi, now assigned to the XIV Gruppo, to regain its forces. There on 20 November it took on force the personnel of the disbanded 201ª Squadriglia Ca.3 of the Regia Marina and its new commander, *cap.* Enrico Cellerino. The need to reorganize and bad weather kept the Squadriglia at Ghedi until 31 December, when it moved to Padua. During 1917 the 9ª Squadriglia had flown 101 war missions. 12 Silver and five Bronze medals were assigned.

In 1918 the unit had on strength Capronis serialled 2333, 4168, 4175, 4179 and 4168. With commander Cellerino there were the pilots *tenenti* Angelo Mamoli and Ettore Trezzi, *s.ten.* Molteni, *sergenti* Savini, Alfredo Fiscali, Arturo Foà, Camillo Rossi, *caporali* Francesco Arbore and Umberto Merlino; the observers *tenenti* Mento, Nicola Discenza and Alfonso Sortino; gunners Ghiretti, Paoletti, Francesco Cossiga, Danilo Pianca and Sinibaldo Santoni.

In January 1918 the unit went against targets in the sector of the Plateaus, now the keystone of the front. On 1 January Ca 4168 (coming from the 201ª) with pilots Rossi and Foà bombed La Comina. On 6 January 4168 with the crew of *serg.* Rossi, *serg.* Foà, *ten.* Sortino and Ghiretti went to bomb the airfield shed at Casarsa, but its central engine failed. They jettisoned the bombs and flew back to Padua to get their bearing. Enemy fire had put their electric system out of work, however, so as they couldn't flash the prescribed signals they were not recognized as friendly, and ground batteries opened up on them. The observer tried in vain to use a hand flashlight, and as nobody turned on the landing lights on the field, the Italians decided to fly to the sea and wait for daylight in order to land. Over Rimini they were again fired at, and they finally managed to land in a field without damaging their Caproni.

On 14 February Ca 4181 of the 10ª Squadriglia crashed at Padua airfield, killing its pilots *ten.* Carlo Stratta and *serg.* Enrico Bino, and also *soldati* Enrico Grazioli and Aristide Degli Esposti, of the 9ª, who had daringly entered the fire trying to save the pilots, and were badly burned, dying in the night: the latter was the brother of Amleto Degli Esposti, a fighter pilot.

During the spring the 9ª Squadriglia, still based at Padua, went on with its usual actions against the enemy in the Trentino sector, and then in June it operated eastwards across the Piave opposing the enemy offensive. In June it received American pilots Lieuts. Arthur M. Beach and William H. Pothoff, in July Wills S. Fitch and William O. Frost. Lieut. Fitch wrote an excellent book "*Wings in the Night*" recalling his experiences as a bomber pilot.

On 1 July 1918 three bombers attacked Austrian barracks behind Mount Meletta. A shell badly damaged the wing of Ca 4175 of *ten.* Carlo Pulcini and *serg.* Rossi, observer *ten.* Filippo La Fiura and *sold.* Luigi Pela, over Rubbio. The pilots jettisoned the bombs and made an emergency landing near Concè, the Caproni was completely wrecked and *ten.* Pulcini and La Fiura were badly wounded. Two Capronis from 9ª Squadriglia took part in the large raid against Pola on 17 July.

On 21 July the 9ª Squadriglia moved to Arquà Petrarca. Its airplanes were Capronis serials 2332, 4083, 4151 and 11495. At this time Mento, now a captain, was commander, with pilots *ten.* Mamoli and Trezzi, Lieuts. Beach, Fitch and

Above: Ca.3 4083 of the 9ª Squadriglia with prominent individual name "*Agla*". (Archive Gianandrea Bussi)

Pothoff, *s.ten.* Giotto Mancinelli Scotti, *m.llo* Empedocle Piccinini and *serg.* Fiscali, Giulio Franchiolo, Merlino, Rossi; observers *ten.* Discenza, Fugalli, Giovanni Vergani and *s.ten.* Vittorio Centa; gunners Pianca, Santoni, Alessio D'Antoni and Dario Mazzarella. Mancinelli and Merlino, with *sold.* Giuseppe Aggiunti, were killed on 5 May 1919 as they were carrying General Milan Stefanik, Minister of War of the Republic of Czecho-Slovakia, from Udine to Bratislava: their Caproni crashed on landing and caught fire.

All through the summer of 1918 the 9ª Squadriglia flew reconnaissance missions, bombed enemy airfields and dropped leaflets and newspapers and, occasionally, baskets of food for the population in the occupied zone. On 28 July during a mission against La Comina airfield, the engine of a Caproni caught fire, and gunner Mazzarella walked on the wing and put out the fire. Lieut. Frost flew two combat sorties: on 11 August against the airfield at Feltre and on 7 September against Belluno, with Lieut. Beach.

On 15 September the unit left its XIV Gruppo and moved to Marcon, now assigned to the Gruppo I. Its letter I stood for Information and this group, with commander the veteran *magg.* Luigi Carnevale, had the task of supplying with food, homing pigeons and eventually also weapons the agents flown behind the enemy lines using Voisins and Macchi L.3's. Their missions departed from nearby Ca' Tessera, the airplanes first flying there from Marcon. Its commander was *cap.* Pasquale Mento, its pilots were *tenenti* Edmondo Marcolin, Mamoli, Trezzi, *m.llo* Piccinini, *sergenti* Fiscali, Franchiolo, Merlino, Primo Pansolini, Rossi, its observers were *tenenti* Centa, Discenza, Fugalli and Vergani; its gunners were D'Antoni, Mazzarella, Pianca and Santoni.

They flew their first such mission on 15 September, on Ca 4193, pilots *ten.* Giuriati and *serg.* Vischioni, observer *ten.* Gallina and gunner *ten.* Trentin dropping pigeons, mission repeated on 16 September, with *s.ten.* Alfredo Palmieri and *serg.* Pansolini with *cap.* Mento and *sold.* Pianca on Caproni 2342, while the crew of *ten.* Trezzi, *serg.* Fiscali, *ten.* Fugalli and *sold.* Mazzarella on Ca 4083 dropped food. Ca 2342 was wrecked on 20 September during a take-off from Marcon on account of an engine failure, injuring *s.ten.* Palmieri and the observer *s.ten.* Domenico Donadelli. Two similar missions were flown in the night of 21 September, with Ca 4193 of pilots *ten.* Marcolin and *serg.* Rossi, and Ca 4083 of *ten.* Trezzi and *serg.* Fiscali.

Activities were curtailed by bad weather in October. On 31 October the squadron was assigned to the Massa da Bombardamento for the final offensive, it returned to Padua airfield, and then back to Arquà Petrarca with the XIV Gruppo. At the end of the war the 9ª Squadriglia had on charge Capronis 4083, 4192 and 4193.

The 9ª Squadriglia was disbanded on 5 December 1918. It had flown 197 combat sorties, earning in 1918 five Silver and ten Bronze medals.

Above: Ca.3 2387 of the 9ª Squadriglia survived the war, here it is at Ciampino, in pristine conditions, almost ten years after its first flight.

Casualties and Combat Losses of the 9ª Squadriglia

Date	Place	Aircraft	Crew	Cause
27.9.1917	Padua	Ca 2327	*Cap.* Jussa KIC, *s.ten.* Sargian WIC, *ten.* Fugalli WIC, *sold.* Ferraris WIC	Accident
29.9.1917	Pola	Ca 4083	*Ten.* De Lorenzi UNH, *serg.* Savini UNH, *ten.* Mento UNH, *ten.* Mento UNH, *sold.* Paoletti UNH	Ground fire & fighter
1.7.1918	Trentino	Ca 4175	*Ten.* Pulcini WIA, *serg.* Rossi UNH, *ten.* La Fiura WIA, *sold.* Pela UNH	Ground fire

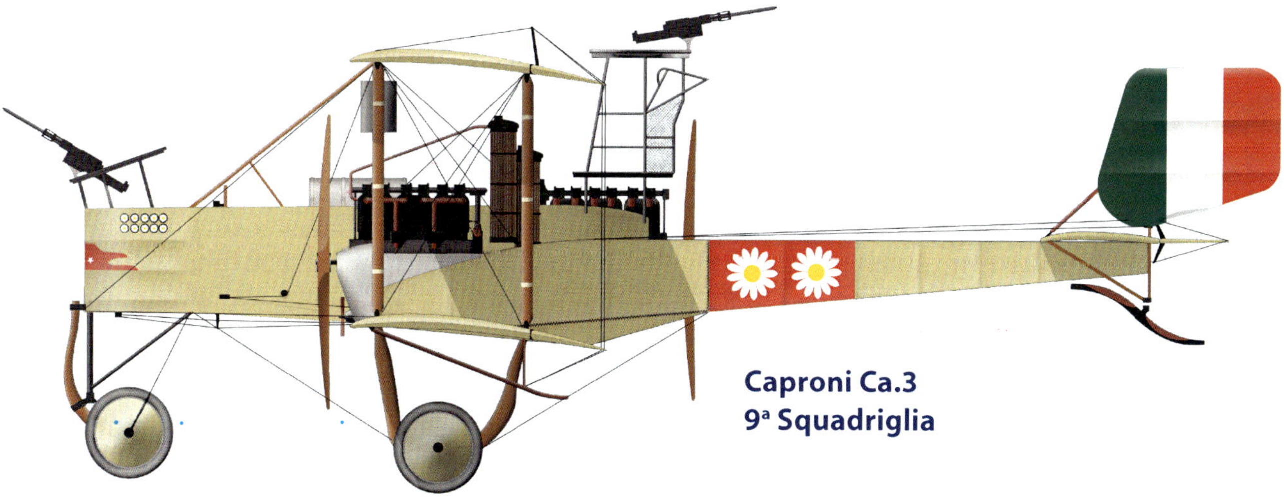

Caproni Ca.3
9ª Squadriglia

10ª Squadriglia

Above: The insignia of the 10ª Squadriglia.

Above: Ca.3 4181 with an unusual arrangement of a windscreen in front of the circular gun ring.

The 10ª Squadriglia was formed on 25 May 1916 at the airfield of Villaverla and when ready it was assigned to Padua in the III Gruppo, operating for the 1ª Armata on the northern front. Its commander was marquis *cap.* Renato De Riso with the pilots *ten.* Filippo Valdimiro, *s.ten.* Ernesto Sattanino, *s.ten.* Giovanbattista Strata, *cpr.* Antonio De Luca, *brig.* Giuseppe Bigliani, *soldati* Romolo Bonacini and Giovanni Facta; the observers *ten.* Guido Pacinotti and *ten.* Vittorio Porcelli; the gunners Carlo Borio, Battista Gays and Dario Nascimbeni. Its Capronis had serials 1120, 1125 and 1126.

On 4 June the 10ª Squadriglia was re-assigned to the IV Gruppo. War operations began on 10 June, when the crew of *cap.* De Riso and *sold.* Bonacini, *ten.* Pacinotti and Porcelli, bombed enemy encampments in the Trentino sector. In June the unit was in action in the same area attacking targets at Mattarello, Tavo, Campomulo and Monte Rovere. On 15 June a biplane attacked the Caproni of *s.ten.* Sattanino, *cpr.* De Luca and *ten.* Pacinotti after a mission on Mattarello. The two airplanes exchanged fire and the Austrian was seen to drop out of the fight diving toward Trento. Brandenburg C.1 61.21 of Flik 7, with pilot *Zugsf.* Karl Rose wounded, made a forced landing. Four days later that same crew had a similar combat, and also this time the Austrian attacker quit the field.

On 29 June, Caproni 1120, with pilots *s.ten.* Sattanino and *sold.* Facta, *ten.* Pacinotti and Borio, had a combat with Brandenburg C.1 26.09 of Heinrich Kostrba and observer Johann Frint, who identified the Italian bomber as a "*large Farman*". The Caproni was damaged, managed to cross the lines and crashed at Zocchi, near Asiago, with the death of Pacinotti, Borio and Facta. The latter was the son of Luigi Facta, a minister, and the last prime minister of Italy before the dictatorship of Mussolini.

On 30 June 1916 the 10ª Squadriglia was sent to Campoformido, still assigned to the IV Gruppo. On 11 July the bomber of *ten.* Valdimiro and *brig.* Bigliani, with gunner Aroldo Moretto, got lost in the fog after a mission and made an emergency landing on the parade grounds of Vicenza: the airplane was damaged but the crew were unhurt. On 1 August three bombers, probably the ones with serials 1125, 1126 and 2405, took part in the mass action against the torpedo works at Fiume. Three bombers also took part in the

Above: The markings for the 10ª Squadriglia in 1917, yellow circles with blue borders. The airplane is probably Ca.3 2329. (Archive Roberto Bassi)

mass action of 13 September against the Lloyd arsenal at Trieste. Enemy seaplanes attacked them, but the Capronis of the 10ª returned safely. Overall, in 1916 the 10ª Squadriglia put up 45 combat sorties.

By 1 January 1917 this unit, still at Campoformido with *cap.* De Riso as commander, had Capronis 1152, 1166 and 1180. Its strength consisted of pilots *brig.* Bigliani, Bonacini and De Luca (now *sergenti* both), *ten.* Mario Panciera di Zoppola, *s.tenenti* Attilio Mazzei, Strata, Adriano Villa and *serg.* Battista Origgi; observers *tenenti* Porcelli, Michelangiolo Caetani and Livio Del Signore; gunners Gays, Moretto, Nascimbeni and Mario Tanzi.

In April and May 1917, the 10ª Squadriglia was active in the Carso sector bombing Chiapovano, Prvacina and other localities, supporting the 10th offensive of the Isonzo. On 19 May Ca 2312 of *ten.* Strata and *brig.* Bigliani suffered the attack of eight Austrian airplanes but survived. On 23 May the Caproni of *ten.* Panciera di Zoppola, *ten.* Alessando Sardi and *serg.* Origgi force landed after an attack by two Brandenburgs of FliegerGeschwader 1.

In June this unit was flying Capronis 2312, 2313, 2370 and 2373. It had basically the same personnel, plus gunner Enrico Ducci. In this period, the unit painted on the fuselages of its Capronis its new insignia, yellow circles with blue border, from one to four to identify each plane in the squadron.

On 2 July De Riso was assigned to the Comando Raggruppamento Squadriglie da Bombardamento (bomber command) and *ten.* Giuseppe Gianolio was acting commander until 6 July when *magg.* Domenico Mondelli replaced him. However, Mondelli was soon posted as acting commander of the XI Gruppo, so command returned to Gianolio on 14 July.

In the night between 27 and 28 July Gianolio was killed together with *ten.* Del Signore as Caproni 2313 crashed returning from a bombing mission to Sesana with 2370 and 2373. As clouds hid the moon, the crew got lost in the darkness, jettisoned the bombs and then tried an emergency landing near the airship sheds at Casarsa, but the airplane hit a power line, *serg.* Bonacini was the only survivor. *Ten.* Panciera di Zoppola then became acting commander. At this time the pilots were *ten.* Panciera, Strata and Mazzei, *s.ten.* Villa, *sergenti* Bonacini, De Luca, Origgi, *brig.* Bigliani, observers were *ten.* Livio Del Signore and Giuseppe Cipriani, gunners *soldati* Natale Coccè, Enrico Ducci, Battista Gays, Antonio Giannotti, Aroldo Moretto, Dario Nascimbeni, Mario Tanzi.

On 6 August 1917, the 10ª Squadriglia formed with 5ª and 9ª a temporary autonomous bomber detachment. On 9 August 1917 the squadron attacked the naval base of Pola by night. Bonacini escaped from hospital in order to take part in this mission, and he got a promotion to *aiutante di battaglia* for his courage. In mid-August the 10ª Squadriglia supported the 11th Offensive of the Isonzo bombing camps and barracks at Gabrovizza, Chiapovano and on the Hermada hill. In this period, the unit had among its observers *ten.* Umberto Nistri, later a pioneer of air photo survey. On 12 August 10ª Squadriglia attacked Chiapovano twice, with six sorties in total.

On 18 August, Caproni 4054 of *ten.* Mazzei, *brig.* Bigliani, gunners Tanzi and Moretto bombed Vojscica and strafed enemy trenches. A fighter attacked them, wounding gunner Tanzi, but the Italian bomber returned to base. The next day the same bomber, crewed by *s.ten.* Strata and De

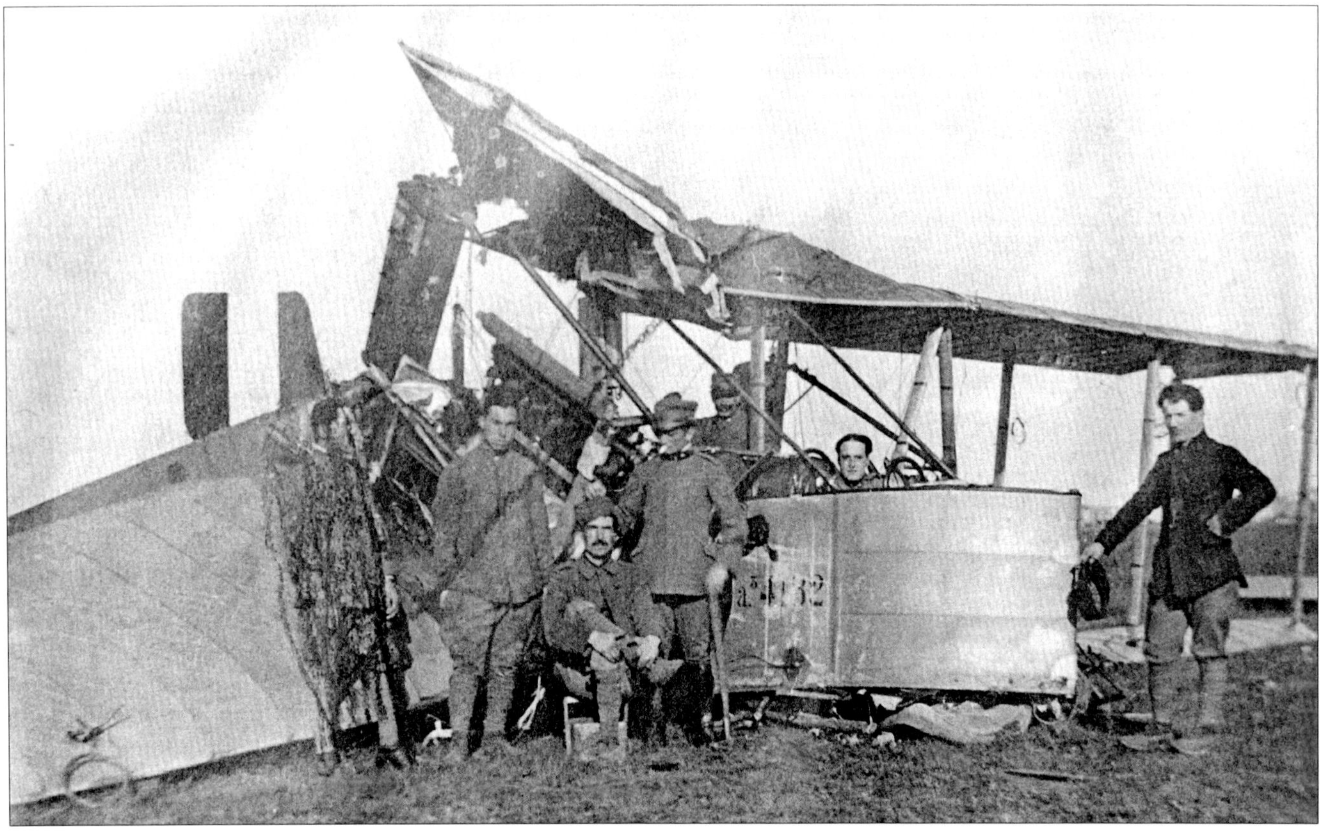

Above: Caproni Ca.3 4182 of the 10ª Squadriglia, that crash landed on 16 December 1917.

Luca, (now an *aspirante*), with Nascimbeni and observer *ten.* Pio Massignani bombed Cerovlje and faced an enemy aircraft. The Austrian oberver saluted them before engaging combat. The Caproni had its central engine out of use, but managed a safe landing at Campoformido. On 3 September its bombers raided again Pola.

On 26 September two fighters attacked the Caproni of *ten.* Luigi Garrone and *serg.* Rosario Zingales near Grahovo, but Italian fighters drove them off. In the night of 29 September Ca 2370 of pilots De Luca and *cpr.* Sebastiano Sandran, observer *ten.* Tommaso Silenzi and gunner Giovanni Grassano failed to return from a mission to Pola. Its fate was only known after the war: Lt. Anton von Nostitz-Rienek, flying seaplane K184, had shot it down. The Caproni caught fire and crashed in the sea. An Austrian torpedo boat picked up the bodies of Sandran and Silenzi (believed to be the only occupants) who were buried with military honors.

At the end of September Panciera di Zoppola, Origgi, Ducci and *ten.* observer Mariano D'Ayala Godoy were assigned to the Distaccamento A.R.

In early October the 10ª Squadriglia, based at Campoformido, had Capronis 2329, 2405, 4065 and 4074. Its commander was *ten.* Massignani, with pilots *brig.* Bigliani, *ten.* Garrone and Mazzei, *serg.* Zingales and Antonio Porta; observers *ten.* Gino Fraschi and Silvio La Guardia; gunners Moretto, Nascimbeni and Natale Coccè.

On 26 October, in the hectic days of Caporetto, the Caproni of Garrone, Zingales, Fraschi and Coccè bombed Tolmino. Having dropped its bombs, it met four enemy fighters. After a ten-minutes combat, one of the assailants was hit and crashed to the ground out of control. The enemy was possibly a German Albatros D.V of *Jasta* 31, its pilot Fritz Jakobsen was wounded but survived.

On 28 October the squadron was moved back to Marcon, leaving behind and burning its Caproni 4074, on 2 November it retreated to Ghedi, with *cap.* Cesare Dima as acting commander, assigned to the XIV Gruppo. During the retreat Ca 2405 was lost as it ran out of fuel, making a forced landing in a rice paddy field with no damage to either crew or plane.

On 15 November, Ca 4169 of Garrone and Zingales with gunners Coccè and Brenno Manfredi flew in a formation attacking the railroad at Levico and barracks near Caldonazzo. The leader had to turn back, so Ca 4169 remained as the leader with two other bombers. Near Caldonazzo five Austrian fighters attacked them; one of them was Albatros D.III 153.36 of the Austrian ace Julius Arigi, whose fire badly wounded Coccè in the back turret. The Caproni, with defensive fire from the front gun only, managed to complete its bombing mission. Then it was damaged by ground fire and, with Arigi still in pursuit, it managed to cross the lines and made a crash landing near Montecchio Vicentino. Coccè was rushed to a hospital but he died of his wounds. Arigi, later on, together with Josef Kiss and Josef von Maier, also shot down Caproni 4191 of the 2ª Squadriglia of Lisa and Bassi.

Above & Right: The Caproni that crashed on 2 February 1918 wiith the death of *magg.* Salomone, *ten.* D'Ayala Godoy, and *serg.* Porta.

On 20 November 1917, *cap.* Lorenzo Crescenzi became the new commander of the 10ª Squadriglia, which on 3 December moved from Ghedi to Padua, where the next day *cap.* Diego Sabbatini took over command.

On 16 December Ca 4182 took off from Padua with other bombers to attack targets in the Cismon area. Its crew consisted of *ten.* Mario Leonardi and *serg.* Giovanni Tasso with gunners Arborige Vidan and Romeo Martinelli. They reached their target and dropped their bombs, but only the ones on the left fell, the five bombs on the right remained attached to the failed bomb carrier. The crew circled the target trying to make the bombs fall down, but meanwhile ground cannons targeted them and a shell broke the fuel lines and damaged the stabilizer. With the left engine dead and gasoline pouring out, the Caproni glided back to the Italian lines and crash landed at Nove di Bassano airfield. During 1917 the 10ª Squadriglia had flown 141 war missions.

Above: Ca.3 4218 of the 10ª Squadriglia in dark night finish. The serial number is painted in very small white characters below the bomb sight.

On 1 January 1918 the 10ª Squadriglia at Padua, part of XIV Gruppo, had Capronis 4065, 4218, 4220, 4227 and 4230. The commander was *cap.* Sabbatini, with pilots *tenenti* Garrone, Leonardi, Mazzei, Panciera di Zoppola, *sergenti* Origgi, Porta, Tasso, Zingales, *brig.* Bigliani and *cpr.* Angelo Pinzuti; observers *tenenti* D'Ayala Godoy, Fiaschi and Massignani; gunners Martinelli, Moretto and Silvio Piovesan. The personnel of the Command of the XIV Gruppo was also attached to this squadron: *magg.* Oreste Salomone, the first Italian airman decorated with the Gold Medal, and the observers *cap.* Angelo Amati, *cap.* Carlo Marchisio, *cap.* Giuseppe Rossi, *cap.* Enrico Ricco Barberis and *ten.* Weiss Poccetti.

In the first two months of 1918 the 10ª Squadriglia took part in actions against the enemy airfields and on the sector of the Asiago plateau, the main threat to the Italian defense system. On 12 January three fighters attacked Ca 4227 of Crescenzi and Leonardi with *magg.* Salomone, but were driven off. On 2 February *magg.* Salomone was killed. Coming back from a night bombing against Campo Rovere, the first one in a moonless night, Caproni 4230 hit a tree when landing at Padua. With Salomone, also *serg.* Porta and *ten.* D'Ayala Godoy were killed, and only the gunner Piovesan survived. On 14 February there was another accident, as Caproni 4181 crashed while taking off, with the death of *ten.* Carlo Stratta and *serg.* Enrico Bini. Soldier Aristide Degli Esposti, of the 9ª Squadriglia, brother of the fighter pilot Amleto, lost his life as he generously tried to save the trapped airmen from the burning wreck.

During the mid-June Austrian offensive, the unit was in action against the Austrian bridges across the Piave River. On 19 June the American pilots Lieuts. Norton Downs Jr., Willis S. Fitch, Paton MacGilvary and Warren S. Wilson arrived. Lt. Downs in September moved to England and died in a flight accident on 23 October 1918. At the end of June *cap.* Sabbatini left, and *ten.* Leonardi became acting commander. In July the unit attacked the airfields of Godega di Sant'Urbano and Portobuffolè and dropped propaganda leaflets. On 17 July two of its airplanes took part in the large daytime raid against Pola.

On 21 July 1918, the 10ª Squadriglia, with the other squadrons of XIV Gruppo, moved from Padua to Arquà Petrarca. At this date the unit had Capronis 4065, 4182, 4188, 4218 and 11489. Its force included pilots *ten.* Leonardi, Lieuts. Downs, Mac Gilvary, Wilson, *tenenti* Angelo Pizzini and Francesco Succi, *sergenti* Angelo Pinzuti, Giannino Solari, Ettore Stucchi and Arturo Zecchi; observers: *tenenti* Cesare Monti, Elio Turchetti, *s.tenenti* Emanuele Capriolo, Noè Fantini and Renato Fattori; gunners Manfredi, Moretto, Vidan, Angelo Monghi, Mario Mugnaini and Giovanni Cerrato.

Right: Caproni 4054 was used after the war for liaison flights, here it is pictured at Cheb airfield, western Czecho-Slovakia, in 1919. (Archive Jiri Railich)

On 30 July five fighters attacked the Caproni of Stucchi, Fitch, Fantini and Moretto that had bombed Portobuffolè. The Italian crew claimed one enemy shot down, then escort fighters drove away the attackers. On 16 August Caproni 4188 and 4218 took off for a leaflet drop mission with five escort fighters. Before Conegliano, accurate ground fire hit Caproni 4218, that was seen going down. The fate of its crew, *ten.* Pizzini, *serg.* Zecchi, *ten.* Monti and *sold.* Cerrato, was revealed by a message dropped by the Austrians: they were all dead, buried with military honors.

In autumn the 10ª Squadriglia was engaged first preparing and then supporting the final Italian offensive, attacking installations at Fonzaso, Sacile, in Val Cismon and the bridges across the Livenza. The final mission of the war, on 1 November, was against the bridges. Meanwhile, on 25 October, *cap. pilota* Casimiro Buttini, *Medaglia d'Oro* holder, had become squadron commander.

At the end of the war this unit was equipped with Caproni 4171, 4227, 4188 and 11489. It included pilots Lieuts. Arthur F. Clement, William O. Frost, George C. Hering Jr., who had arrived in October, *ten.* Giovanni Lodi, sergenti Solari, Stucchi, Tasso, and *m.llo* Luigi Monnier; observers *ten.* Fantini, *ten.* Giorgio Bruno and *s.ten.* Giuseppe Panselli; gunners Manfredi, Luigi Assier, Pietro Gobbi, Giuseppe Pezzaia and Giuseppe Zamolo. In 1918 it had flown 141 combat sorties, 296 in the whole war.

The 10ª Squadriglia was disbanded on 6 February 1919. Its insignia, a four-leaf clover, is now part of the current shield of the Aeronautica Militare, representing bomber units.

Casualties and Combat Losses of the 10ª Squadriglia

Date	Place	Aircraft	Crew	Cause
29.6.191	Cima 12	Ca 1120	*S.ten.* Sattanino WIA, *sold.* Facta KIA, *ten.* Pacinotti KIA, *sold.* Borio KIA	Fighters
23.5.1917	Carso	Caproni	*Ten.* Panciera di Zoppola UNH, *Serg.* Origgi UNH, *ten.* Sardi UNH	Fighters
28.7.1917	Casarsa	Ca 2313	*Ten.* Gianolio KIC, *serg.* Bonacini WIC, *ten.* Del Signore KIC	Accident
29.9.1917	Pola	Ca 2370	*Asp.* De Luca KIA, *cpr.* Sandran KIA, *ten.* Silenzi KIA, *sold.* Grassano KIA	Fighters
15.11.1917	Trentino	Ca 4169	*Ten.* Garrone UNH, *serg.* Zingales UNH, *sold.* Manfredi UH, *sold.* Coccè KIA	Fighters
16.12.1917	Cismon	Ca 4182	*Ten.* Leonardi UNH, *serg.* Tasso UNH, *Serg.* Vidan UNH, *sold.* Martinelli UNH	Ground fire
2.2.1918	Padua	Ca 4230	*Magg.* Salomone KIC, *serg.* Porta KIC, *ten.* D'Ayala KIC, *serg.* Piovesan WIC	Accident
14.2.1918	Padua	Ca 4181	*Ten.* Stratta KIC, *serg.* Bini KIC	Accident
16.8.1918	Susegana	Ca 4218	*Ten.* Pizzini KIA, *serg.* Zecchi KIA, *ten.* Monti KIA, *sold.* Cerrato KIA	Ground fire

11ª Squadriglia

Above: Caproni Ca.3 2395 of the 11ª Squadriglia in front of the big airship hangar at Grottaglie airfield, where Boeing 787 Dreamliner fuselages are now built.

The Ministry of War decided in April 1916 to form a new Caproni squadron to be assigned to Albania, where Italian troops the XVI Army Corps with main base at Valona harbor were facing the Austrians in the north, who in February 1916 had conquered Durazzo (now Durres). The manufacturer had the bombers ready by June, but the airfield near Valona was not, so, with the permission of the Regia Marina, the Capronis were sent by railway to Grottaglie, Southern Italy, to be housed in the local empty airship hangar. When the pilots arrived, the 11ª Squadriglia was formally activated, on 7 June 1916, at the command of *cap.* Ercole Ercole, with five bombers.

Works at Tahyraqua airfield (also written Tahiraga, a place named after a local family, near Valona) progressed slowly, so on 25 July two bombers took off from Grottaglie to attack Durazzo, carrying two Navy observes, STV Caffaratti and Spagnolo, flying for over 5 hours above the sea, with the escort of flying boats from Brindisi, that earned the praise of the Chief of Naval Staff. On 9 August the observer *cap.* Renato Pascale reported from Tahyraqua that the air base was still deficient, with no provision for lodgings and a runway full of holes. Finally on 25 August three bombers left Grottaglie and deployed to their new base, with two more Capronis waiting in the airship base for lack of pilots. These bombers were then left in Italy as a reserve, and moved on to Foggia when the hangar was returned to the Navy, that had to house one of its airships. Some sources report that Austrian ace Julius Arigi on 18 September 1916 shot down a Caproni in Albania, but no Italian confirmation has been found.

The 11ª Squadriglia incurred its first loss on 12 October when a Caproni with pilots *cap.* Ercole and *brig.* Albino Mocellin and as observer *cap.* Emilio Corbelli met Brandenburg C.I 68.04 of *Offstvtr.* Josef Siegel near Zarnes. The Austrian pilot killed Mocellin and Corbelli and riddled the fuel tanks of the bomber. *Cap.* Ercole force landed, on the ground, then he burned the wreck with the bodies of his two comrades and escaped, fighting his way back through 50 kilometers of enemy territory, and seven days later he reached the Italian positions. For this feat Ercole was awarded the *Medaglia d'Oro al Valor Militare.*

In October a hurricane wrecked the hangar and all the airplanes were damaged, so operations were suspended. On 9 December 1916 the 11ª Squadriglia became a component unit of the VIII Gruppo. On 26 December the crew of pilots *s.ten.* Ettore Campioni, *serg.* Giovanni Buffoli and *cap.* Renato Pascale took the commander of Italian troops in Albania, *gen.* Giacinto Ferrero on a long reconnaissance mission all along the frontline. During 1916 11ª Squadriglia had provided 11 reconnaissance missions, two standing patrols, two offensive patrols and seven bombing sorties, for a total of 22 combat flight hours.

On 1 January 1917 the 11ª Squadriglia had as commander *cap.* Pascale, an observer, with pilots *sottotenenti* Campioni,

Above: A Caproni at Tahiraqua. Albanian workmen are still working to improve the runway.

Marcello Lodi, Ivo Oliveti, Giovambattista Pittaluga, *m.llo* Giuseppe Marini, *sergenti* Alfredo Alderighi and Buffoli; observers *ten.* Leonardo Giuliano and *s.ten.* Mario Candela; gunners Serafino Bernocchi, Guglielmo Bragaia, Michele Braida and Pietro Muzzi. After the hurricane, rain kept the airfield unusable, and flights were not resumed until April 1917, when the Caproni 450s replaced the Ca 300s. Meanwhile on 3 February *cap.* Diego Sabbatini replaced Pascale as commander.

The first combat flight of 1917 was a reconnaissance mission of two planes on 27 May. Two days later an airplane fell in the sea after take-off, without casualties. In June, July and August the 11ª Squadriglia attacked Fjeri, Berane, camps at Mount Licovum, at Kuci and the bridge of Metali. On 9 September there was an accident to Ca 2395 "*Audace*" of pilots *ten.* Antonio Magnacavallo and *serg.* Italo Lucchi, observer Giuliano and gunner Lazzaro Santucci. The lucky cat and star painted on the nacelle were successful, because the crew was unhurt.

On 17 September Sabbatini passed his command to *cap.* Attilio Viziano, whom the observer *cap.* Luigi Criniti replaced on 18 November. In the autumn of 1917 the squadron again bombed the bridge of Kuci, Cerevoda, and other enemy targets, the mission of 19 December had the escort of six Nieuports of 85ª and five SAMLs of 116ª Squadriglia.

Caproni 2395 had another failure on 7 October, and once again Lucchi and Magnacavallo nursed it back safely to the airfield, earning the praise of *magg.* Rada, commander of the VIII Gruppo. During the year 1917 the squadron had flown 29 combat sorties.

On 1 January 1918 the 11ª Squadriglia with *cap.* Criniti had the pilots *tenenti* Magnacavallo, Ludovico Sergardi, Giovambattista Trotti, *s.ten.* Gino Santi, *sergenti*, Lucchi, Luigi Barberi, Giovambattista Boero, Emilio Magistrelli and Guido Violi; the observers *ten.* Candela, Giuliano and *s.ten.* Italo Migliasso; gunners Bragaia, Braida, Luigi Bartoletti and Otello Firmani.

In the late winter and spring of 1918, the unit attacked military installations at Liusna, Metali, Vizeza and Cercvodo with three sorties of three planes in April, two on 4 May and five, in two waves, on 16 May. On 25 June it attacked the Austrian airfield at Fjeri. With just two bombers, during the Italian offensive the 11ª Squadriglia flew two missions each day for one week. In the afternoon of 6 July 1918 there was the loss of Caproni 4164 and its complete crew, *ten.* Magnacavallo, *serg.* Boero, *ten.* Candela and *cap.* Giuliano. The bomber, that on its nacelle carried the names "Piave" and "Osum" (the frontline rivers in Italy and in Albania) had bombed Metali with another Caproni and the escort of two British fighters and two fighters of the 85ª Squadriglia. Ca 4164 had left the formation, and when it was almost over its airfield it was attacked. *Oblt.* Bela Macourek, commander of Flik 6, on Aviatik D.1 238.51, shot it down.

Left: *Cap*. Ercole returns to the Italian lines after a 50 km. trek through enemy territory.

After these events the 11ª Squadriglia, now with *ten.* Trotti as acting commander, had only one serviceable airplane, three pilots and one observer. Still, the single plane and crew fought on, flying in combat almost every day in July and August, until a second Caproni arrived, and new men, the pilots *sergenti* Gino Abati, Giuseppe Brambilla, Armando Manicardi and the observer *cap.* Angelo Balestrino. No more combat missions took place after 14 September 1918, because the rapid advance of Italian troops had moved the frontline beyond the range of the airplanes, and also on account of bad weather. Combat sorties in 1918 were 43, bringing the total at 94.

At the end of hostilities, the 11ª Squadriglia was reinforced and remained in Albania to co-operate against the insurgent bands operating in that country until the final retreat of Italian troops in August 1920.

Casualties and Combat Losses of the 11ª Squadriglia

Date	Place	Aircraft	Crew	Cause
12.10.1916	Zarnec	Caproni	*Cap*. Ercole UNH, *brig*. Mocellin KIA, *cap*. Corbelli KIA	Fighters
6.7.1918	Ostima	Ca 4164	*Ten*. Magnacavallo KIA, *serg*. Boero KIA, *cap*. Giuliano KIA, *cap*. Candela KIA	Fighters

Below: This Caproni of the 11ª Squadriglia in Albania was christened "*Falco*", hawk.

Right: A Ca.3 at Tahiraqua airfield.

Above: The Caproni Ca.3 exhibited in the Historical Museum of the Italian Air Force at Vigna di Valle, actually a post-war Ca.3, 23174, was painted in 2006 as Ca 4166 of the 11° Squadriglia. Behind it is the fuselage of a Junkers J4.

12ª Squadriglia

Above: A Caproni of the 12ª Squadriglia at Tripoli.

When the Great War broke out, Italian control over its new colony of Libya, conquered with the war against Türkiye of 1911–1912, was still far from complete. The growing resistance of the local population and the impossibility of sending more troops from the mother country forced the Italians to limit their occupation to a narrow coastal area, with two strongholds at Tripoli and Homs in Tripolitania, the western part, and Derna, Benghazi and Tobruk in Cyrenaica, the eastern part of the country. Türkiye and Germany fostered the revolt of the Libyans, and submarines supplied money and weapons to the rebels, that had capable Turkish officers as advisors.

In the summer of 1916, the local Squadriglia Farman and observation balloons section were to be supplemented with a squadron of Caproni bombers. This unit, the 12ª Squadriglia, was formed on 18 September 1916 at Pordenone (La Comina airfield) and it moved with its four Caproni 300 hp on 2 October, going to Tripoli, at the Mellaha airfield. Its material was temporarily housed in the airship shed, next to the Squadriglia Farman.

The airplanes were assembled and they flew their first combat mission on 26 October 1916, bombing Fonduq el Scerif. On 1 December two bombers effectively attacked the castle and the fort of Azizia, where the rebel chief Suleiman el Baruni had just camped. Exploiting the commotion, *serg. pilota* Raul Moore and the observer *ten.* Ermanno Bartolini, of the 104ª Farman managed to escape and reached Tripoli the next day. They had been captured on 25 October.

In December 1916 the 12ª Squadriglia flew 16 bombing missions, hitting Fonduq Ben Gascir and the oasis of Sorman and Zanzur. It also flew methodical reconnaissance patrols among the oasis of Khaarian, Kallet and Suani el Fandi and along the coast beyond Zuara.

On 1 January 1917 the 12ª Squadriglia at Mellaha had as commander *cap.* Alessandro Raffaelli, with pilots *cap.* Attilio Gelli, *ten.* Gaspare Milanesi, *s.ten.* Felice Marzetti, *m.llo* Silvio Picolla, *serg.* Riccardo Bodo, and *serg.* Giovanni Conti; gunners Emilio Marè, Edmondo Radice and Anacleto Scola.

On 10 January the Capronis took off to support militia troops engaged in a three-hours battle with the rebels. The airplanes first dropped bombs and steel darts, then they strafed the insurgents in the open terrain, who put up strong rifle fire. On 16 January, during a combat mission, a Caproni suffered a broken oil pipe, and *sold.* Marè walked on the wing to fix it, and the bomber returned safely. The same problem happened on 10 February, and this time it was Scola who walked the wing and fixed the engine.

In February and March 1917, despite very bad weather, the Capronis flew 21 bombing and reconnaissance missions, mainly against coastal localities east and west. In March

Above: A local personality is flown on Ca.3 2396 of the 12ª Squadriglia. (Archive Fiorenzo Longhi)

more airplanes arrived from Italy, so the squadron had a force of four bombers in service and two more crated as reserves.

In April 1917 the Capronis were used to set on fire the barley fields providing food for the rebels. Over the fields at Zanzur, Zavia and Borman they dropped a total of 3,600 kg of explosives and 1,270 kg of incendiary liquids. Incendiary bombs proved defective, and Raffaelli with his pilots designed a bomb built with spare containers that proved efficient, so the design was sent to Italy. In April there was a high level of enemy submarine activity along the Libyan coast, and the sloop "*5 ottobre*" and the tug "*Progresso*" were sunk. The Capronis were tasked with maritime patrol flights looking for submarines and for drifting mines. One more new mission was to escort outgoing ships as long as their range permitted it.

On 16, 18 and 19 June the Capronis were particularly successful bombing Azizia, where Suleiman el Baruni was hosting a meeting of rebel chiefs. On 23 August a ship spotted a submarine 10 miles north of Tripoli, a Caproni and a Farman promptly reached it and forced it to submerge. On 30 August again a Caproni attacked a submarine.

At the end of the summer 1917 the 12ª Squadriglia at Mellaha had as commander *cap. pilota* Attilio Gelli. It was equipped with Caproni 300 hp 1210, 1216 and 1224 and its force included pilots *tenenti* Adolfo Crema Bianchi, Nilo Tibaldi, *s.ten.* Umberto Copasso, *sergenti* Bodo, Conti, Milanesi, Carlo Prina, Giovanni Marini and Felice Strinasacchi. The only known gunner is Radice.

The Squadriglia operated in support of the Cassinis column, marching from Zuara toward Tripoli in September, earning the citation of the Governor of Libya, *Gen.* Ameglio. The Capronis were used also as scouts for the mobile units and to carry ice for medical purposes. On 17 September a machine gun pierced the fuel tanks of a Caproni with pilot *s.ten.* Copasso during a bomb raid. Strinasacchi stopped the leaks with his hands and the airplane returned to base.

In the last three months of 1917 flying activities were much reduced on account of heavy rains and of the fatigue of the aircraft. Five Caproni 450 hp had arrived to replace the worn-out 300 hp, and they were assembled, while the old ones were crated and shipped back to Italy. They were not new airplanes, however, but veterans of a long service on the main front, that the manufacturer had overhauled. These Caproni 450 had serials 2315, 2320, 2396, 4070, and 4073.

During 1918 the 12ª Squadriglia continued its usual missions of reconnaissance and control of rebel armed bands.

Above: Ca.3 4073 at Mellaha airfield.

On 20 January the commander, *ten. pilota* Adolfo Bianchi Crema, had a flight accident and was seriously wounded. On 26 August two Capronis and three Farmans bombed Ben Aden.

A major action was planned against Misurata, an enemy-held port that had an important radio station and that was a calling port for German submarines. After careful planning, the mission took place on 8 September, with the participation of British flying-boats and of the Farmans. The commander of an Italian ship reported that fires in Misurata could be spotted from 12 miles away. On 18 September another radio station, at Beni Ulid, 130 km from Tripoli, was also bombed.

At the end of 1918 the unit passed under the command of *cap. pilota* Umberto Bassani. Its personnel consisted of pilots *cap.* Luigi Lanata, *tenenti* Bodo, Copasso, Prina, Nicola Collarile, Ireneo Taverna, *serg.* Antonio Pasta; observers *cap.* Ferruccio Pizzi, *ten.* Stefano Pucci and *ten.* Edmondo Silvestri; gunners Radice, Alberto Pecci and Arrigo Crisanti. The end of the conflict in Europe did not stop the activities of the 12ª Squadriglia, that remained operative in the long post-war conflict for the definitive occupation of Libya, that was completed only in 1931.

Casualties and Combat Losses of the 12ª Squadriglia

Date	Place	Aircraft	Crew	Cause
20.1.1918	Libya	Caproni	*Ten.* Bianchi Crema WIC	Accident

Above: Mellaha was a large, well-organized airfield; in 1937 it was enlarged and called Castel Benito, after Benito Mussolini. After World War 2 the U.S. Air Force used it as Wheelus Air Base, today it is Mitiga International Airport.

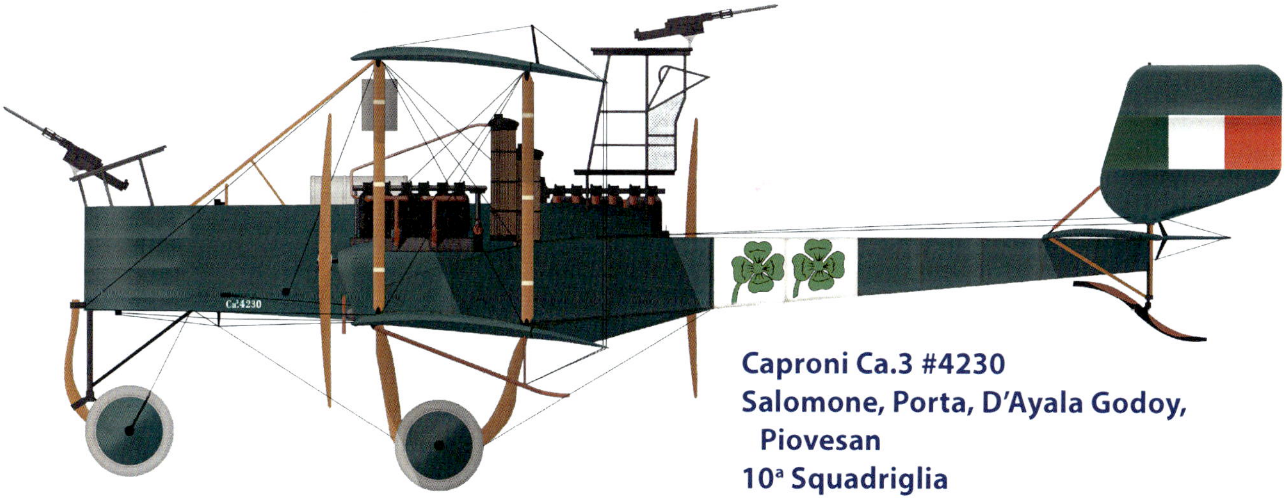

Caproni Ca.3 #4230
Salomone, Porta, D'Ayala Godoy, Piovesan
10ª Squadriglia
February 1918

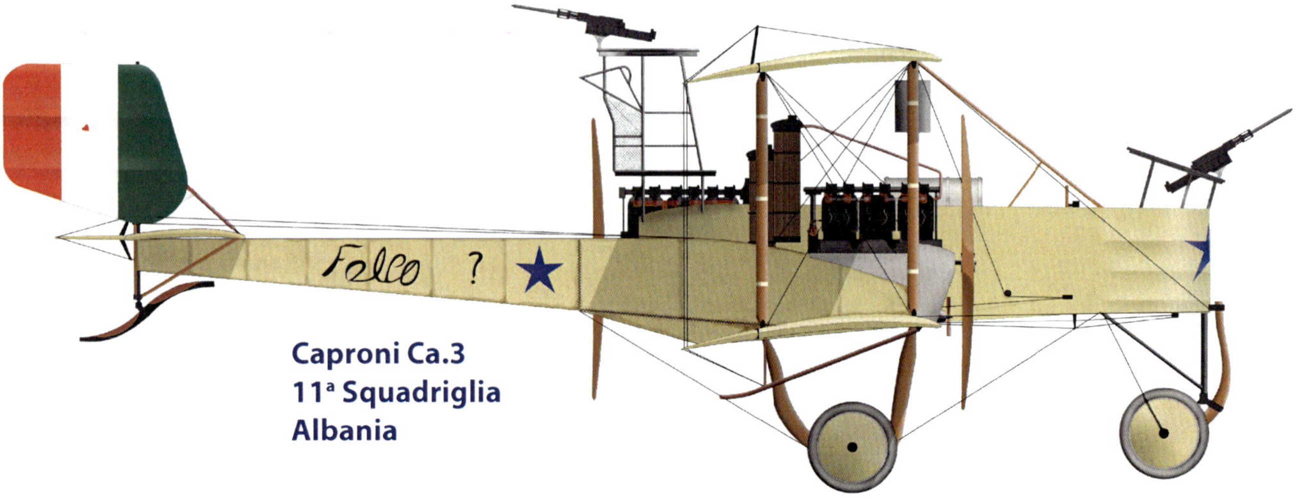

Caproni Ca.3
11ª Squadriglia
Albania

Caproni Ca.3 #2395
11ª Squadriglia
Albania

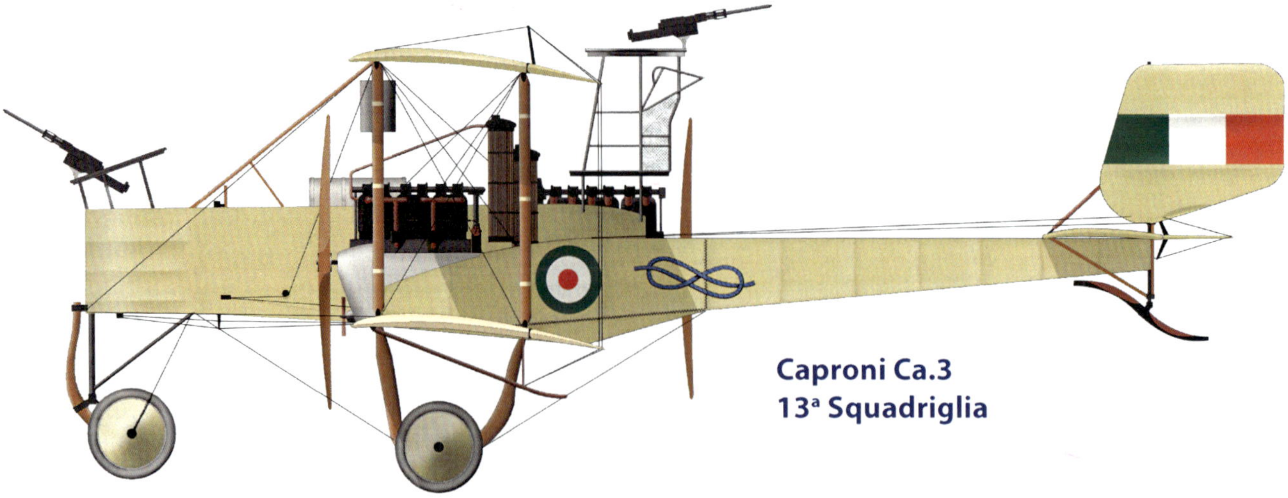

Caproni Ca.3
13ª Squadriglia

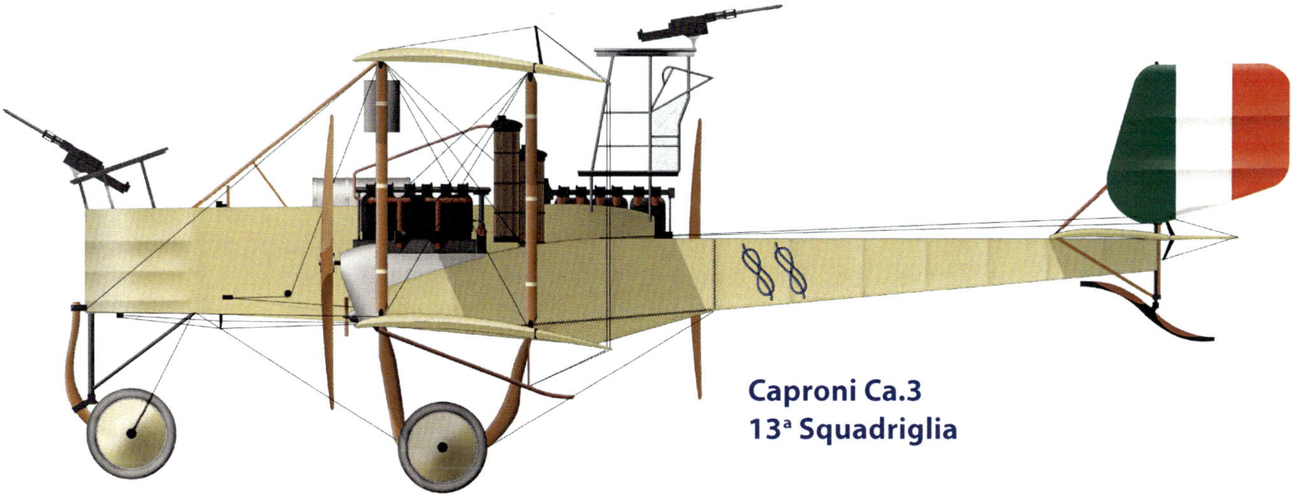

Caproni Ca.3
13ª Squadriglia

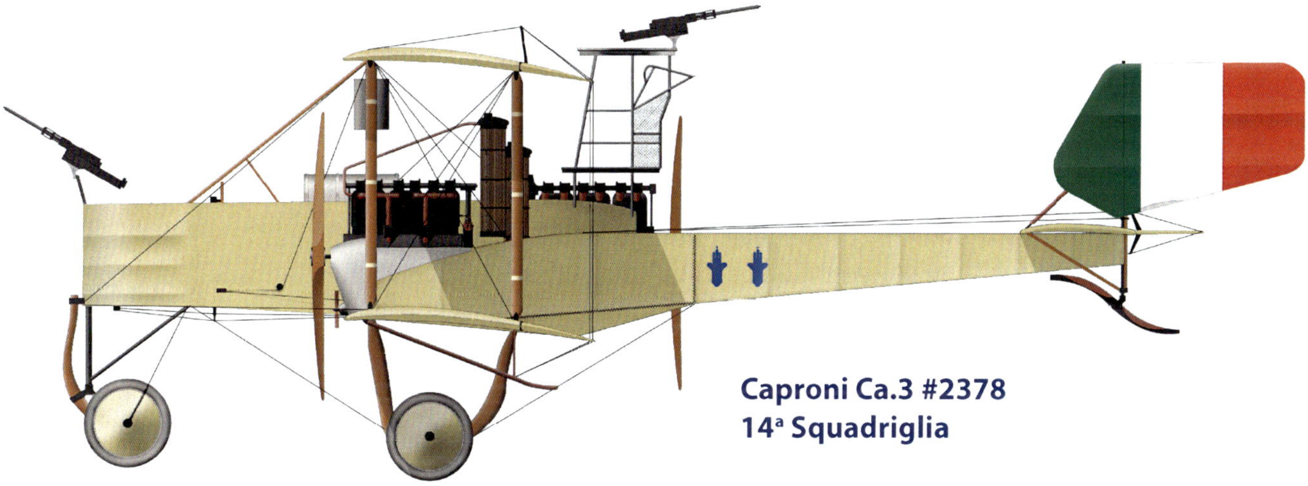

Caproni Ca.3 #2378
14ª Squadriglia

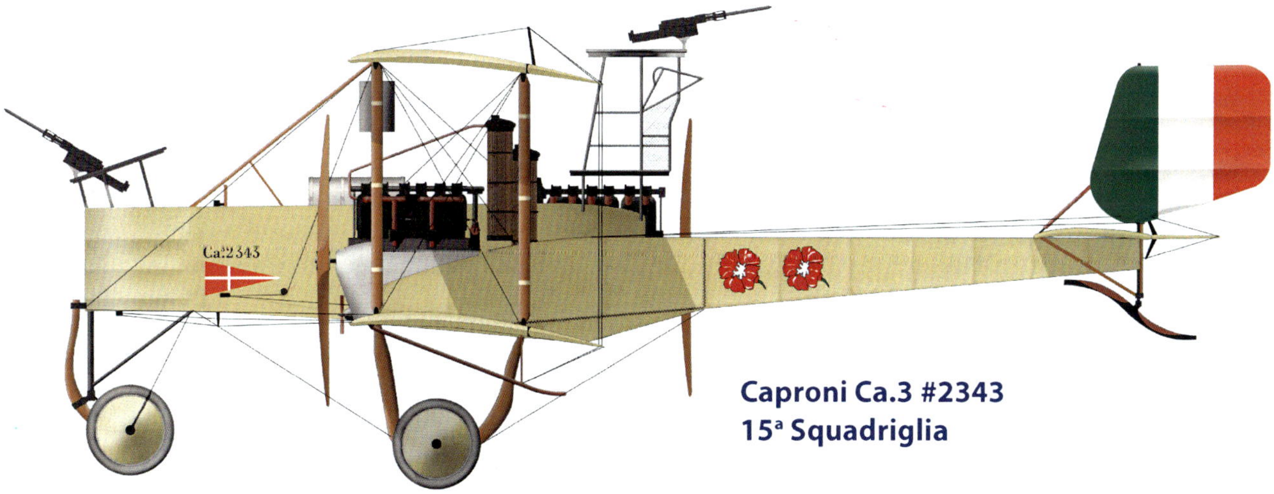

Caproni Ca.3 #2343
15ª Squadriglia

13ª Squadriglia

Above: A Caproni of the 13ª Squadriglia showing the knot of Savoy, an unusual small flag on the rudder, and camouflage with wide bands on the wings and horizontal tail. (Archive Caliaro)

The 13ª Squadriglia was formed in August 1916 at La Comina airfield and it became operative in October. It was assigned to the IV Gruppo, equipped initially with Capronis 300 hp 1186, 1187, 1191 and 2607. Its commander was *cap.* Piero Opizzi, its pilots were *ten.* Federico Zapelloni, *s.tenenti* Luigi Contini, Carlo Frova, Filippo Serafini, *m.llo* Luigi Ponsiglione and *serg.* Antonio Angelelli; its observers were *ten.* Luigi Ricci and *s.ten.* Giuseppe Fantini; its gunners: Angelo Balzarotti, Ernesto Mecci and Italo Simonetti.

The 13ª Squadriglia flew its first war sortie on 3 November 1916, with Caproni 1237 of *s.ten.* Frova flying against the station of Dottogliano. Two Austrian seaplanes attacked it over the target, and anti-aircraft fire was also present: with its central engine damaged, the bomber made a forced landing near Oleis; gunner Simonetti was lightly wounded in a leg, the other crew members were unhurt.

Weather conditions were prohibitive until April 1917, meanwhile, on 10 April, Zapelloni, now a captain, became the commander, as Opizzi became commander of the new XI Gruppo. The first combat sortie was on 20 April, Ca 1186 with pilots Zapelloni and Angelelli. In the evening of 3 May 1917 Ca 2607 took off with pilots *s.ten.* Frova and *ten.* Lorenzo Leveroni and gunner Luigi Caccialanza to bomb the Austrian airfield at Prosecco. A strong headwind slowed them down, and having dropped the bombs, the central engine had to be throttled down, and the pilots made a successful emergency landing near Susegana.

The 13ª Squadriglia was active during the 10th Offensive in May 1917, hitting targets at Vogersko, Bosco and San Daniele. On 23 May Ca 1186 with pilots *cap.* Zapelloni and *serg.* Angelelli, gunners Caccialanza and Otello Firmani, took off at daybreak to bomb targets near Birhula. Five Austrian fighters attacked it after it had dropped its bombs. The pilots flew the big bomber in tight turns and gunner Firmani shot at an enemy fighter, seeing it getting out of the fight in a steep dive. Austrian sources on that day report the loss of Hansa Brandenburg D.I 28.25 of Flik 42/J, with its seriously wounded pilot *Leutnant* Richard Melzer in an emergency landing.

In June 1917 the 13ª Squadriglia started replacing its Ca 300s with the more powerful Ca 3 450 hp having serials 2375, 2376, 2379 and 2380. Personnel at this time, included pilots *tenenti* Zapelloni, Filippo Serafini, Carlo Frova, *s.ten.* Renzo Leveroni, *asp.* Luigi Ponsiglione, *sergenti* Antonio

Above & Below: A Caproni of the 13ª Squadriglia crashed in a marsh, this could be the accident of 26 February 1918.

Angelelli, Benedetto Claris Appiani, *cpr.* Giacinto Gado, and observers *tenenti* Giuseppe Fantini, Luigi Ricci, Pio Massignani, Alberto Scalcerle, Luigi Zocca, gunners *soldati* Luigi Barbisotti, Luigi Bertoletti, Caccialanza, Firmani, Ernesto Macci, Guido Rossi. TV Ernesto Pacchiarotti of the Navy was detached to this unit.

In the nights of 3, 4 and 9 August 1917 the 13ª Squadriglia sent out all its serviceable bombers against the naval base of Pola. The command of the Raggruppamento Squadriglie da Bombardamento commended *cap.* Zapelloni and all the crews taking part in these actions. The 13ª was the only squadron that took part to the three raids with all of its Capronis reaching the target. Besides the moral value, however, real damages inflicted were insignificant. In this period the Squadriglia used Capronis serialled 2379, 2380, 4047, 4055 with as insignia "the knot of Savoy", in blue or red. Ca 2380 also featured the dates and targets of its sorties painted on its central nacelle.

In the summer 1917 the 13ª Squadriglia took part in the 11th Battle of the Isonzo attacking Mount Gabriele, Ternova

Above: Ca.3 2380 being exhibited in an airfield not at the front.

and the railroad at Prosecco. On 20 September 1917 the pilots *ten.* Giovambattista Pittaluga and *ten.* Ivo Oliveti, the observer Scalcerle and gunner Mecci were detached to Taliedo to provide a crew for the Distaccamento A.R. Oliveti, formerly of the 11ª Squadriglia, after the war was an important exponent of the Fascist Party, he died flying in Ethopia in 1936.

Following the rout of Caporetto, the 13ª Squadriglia was active trying to contain the advance of the Austrians and Germans, bombing them at S. Lucia and Ponte Pinzano. On 26 October, third day after the breakthrough, Ca 2379 with the pilots *ten.* Frova and *serg.* Gado and the gunners *sold.* Barbisotti and Caccialanza failed to return from a mission over S. Lucia di Tolmino. The whole crew was killed; Ca 2379 being one of the ten Capronis claimed that day by the German fighters of Jasta 31 and Jasta 39.

As the front fell back, the unit was ordered to move to San Pelagio, and it flew there on 31 October, having Capronis 2376, 2380, 2406, 4046, 4055, 4167, and 4173, leaving behind at La Comina Ca 4047, burned to prevent its falling in enemy hands. On 1 November an enemy airplane welcomed them at San Pelagio, dropping some bombs, hitting the hangar of the squadron and badly wounding the observer *ten.* Luigi Bertani.

On 7 November 1917 six enemy fighters attacked Caproni 4173 crewed by *ten.* Oliveti and *cpr.* Giovanni Conti with gunners Mecci and Firmani and prevented it from completing its mission against Meduna. The fighters pursued the formation of three bombers well into the Italian lines and Firmani saw one of them going down trailing smoke. When they landed at Arcade the other crews confirmed the gunner's claim. On 16 November the 13ª Squadriglia turned in its unserviceable aircraft and was equipped with Caproni 2336, 2380, 4046 and 4167

On 1 January 1918 the 13ª Squadriglia had commander *cap.* Zapelloni, its pilots were *capitani* Leveroni, Vittorio Bergonzi, Cesare Peraglie, *tenenti* Oliveti, Pittaluga, Serafini, *sergenti* Angelelli, Claris Appiani, Conti, Giovanni Magnino, *cpr.* Giovanni Macri and *cpr.* Giuseppe Federici; its observers were *tenenti* Ricci, Scalcerle, Giorgio Morigi, Pietro Rocca, Ugo Serani and *s.ten.* Luigi Bertani; its gunners: Mecci, Bertoletti, Firmani, Ferdinando Mattioli and Antonio Zancan.

On 5 January Ca 2336 suffered an engine failure and force landed, being badly damaged, while its crew was unhurt. On 24 January the Squadriglia began modifying its bombers, fitting additional fuel tanks for a projected raid against Bolzano, that, however, did not take place. This activity did not stop the usual missions, and the unit in that month bombed Levico, Primolano and the airfield at La Comina, now used by the enemy.

On 20 February Zapelloni passed his command to *cap. pilota* Lorenzo Leveroni and was assigned to the Bomber Command. In 1922 he was awarded the *Medaglia d'Oro* which, together with his six Silver Medals, makes him the most decorated Caproni pilot. On the 26th there was another loss, after a night sortie against the railroad station at Bolzano. With its central engine wrecked by anti-aircraft fire Ca 11488 of *ten.* Pittaluga and *serg.* Conti, *ten.* Ricci and *sold.* Firmani, got lost in the fog and found itself over Contarina, near Rovigo. The pilots tried to land in an open field, that actually was a marsh, where the bomber turned over. Pittaluga, a veteran of the raid against Cattaro, was trapped beneath the wing and drowned.

On 16 March Caproni 4223 of the 1ª Squadriglia crashed

Above: Federico Zapelloni in front of his Ca 2380 showing the ribbons of his decorations and the names and dates of the bombed targets. (Archive Longhi)

in a training flight, killing its crew of *serg. pilota* Giuseppe Modena and gunners Ugo Campi and Emidio Bonora, and also *s.ten.* Giuseppe Federici of the 13ª, plus one civilian. Two days later in a similar accident *cap.* Peraglie was also injured.

On 1 June 1918 the 13ª Squadriglia had this structure: commander *cap.* Leveroni, pilots *ten.* Oliveti and Orazio Priore, *asp.* Eugenio Piccone, *sergenti* Antonio Arnoldi, Conti, Macri, Magnino, and Antonio Rossini; observers *cap.*

Above: Ca.3 11494 flew with the 13ª Squadriglia. (Archive Apostolo)

Enrico Leonetti, *tenenti* Scalcerle, Giuseppe Cibella, Rosario Contarini and Luigi Ricci, *s.ten.* Serani; gunners: Bertoletti, Mattioli, Mecci, Guido Aldrovandi, Raffele Aleiback, Michele Rapetti, Francesco Sernia and Zancan. The unit then had four Capronis, among them serials 4046, 4142, 4147.

During the Austrian offensive of mid-June the 13ª Squadriglia bombed troops at S. Donà and Falzè. On 15 June Ca 4142 coming back from a mission turned over and was wrecked. *Asp.* Piccone and *serg.* Magnino were slightly injured, while Serani and Sernia were unhurt. The crews tested the new and unpopular Caproni Ca.5.

Il 20 June the 13ª Squadriglia received the American Lieuts. Allen Williams Bevin and John Park and some days later Lieut. Oble Hugh Mitchell. On 7 July enemy artillery hit Caproni 11494, (pilots Priore and Magnino, gunner Mattioli) and the observer *ten.* Serani was seriously wounded, but he completed the mission. The same bomber, crewed by Lieut. Park and *serg.* Magnino, *ten.* Contarini and *sold.* Mattioli, on 11 August suffered engine troubles, jettisoned its bombs over Arten and lost height over Grappa Mountain, where soldiers shot at it with their rifles, badly wounding *ten.* Contarini. The pilots landed immediately at San Pietro in Gu to get medical care for their observer, but it was too late, and Contarini died of his wound the next day.

On 16 September 1918 *cap. pilota* Diego Sabbatini took over command of the 13ª from Leveroni.

At the end of the war the 13ª Squadriglia was flying Capronis 2385, 4167, 4173, 4187 and 11494. Its pilots were Rossini, *tenenti* Priore and Giuseppe Li Donni, Lieuts. Park, Mitchell and Emory E. Watchorn, *sergenti.* Arnoldi, Conti, Magnino and Rossini; its observers were *cap.* Leonetti, *tenenti* Cibella, Scarcerle, Serani and *s.ten.* Ottavio Petri; its gunners: Aldrovandi, Aleiback, Rapetti, Sernia, Zancan, Umberto Rosa and Arturo Vaiani. During the war it had flown 260 combat sorties, 103 of them in 1918.

After the end of the European war this unit was moved to Libya, with commander *cap.* Francesco Lanzafame. It returned to Italy on 2 January 1920 and then it was disbanded.

Above: A Caproni showing two Exclamation marks. There was another one showing two Question marks, but to which squadron they belonged is not known.

Casualties and Combat Losses of the 13ª Squadriglia

Date	Place	Aircraft	Crew	Cause
3.11.1916	Dottogliano	Ca 1237	*S.ten.* Frova UNH, *sold.* Simonetti WIA	Fighters
26.10.1917	S. Lucia	Ca 2379	*Ten.* Frova KIA, *serg.* Gado KIA, *sold.* Caccialanza KIA, *sold.* Barbisotti KIA	Fighters
26.2.1918	Contarina	Ca 11488	*Ten.* Pittaluga KIA, *serg.* Conti UNH, *ten.* Ricci UNH, *sold.* Firmani UNH	Ground fire
15.6.1918	San Pelagio	Ca 4142	*Asp.* Piccone WIC, *serg.* Magnino WIC, *ten.* Serani UNH, *sold.* Sernia UNH	Accident
7.7.1918	?	Ca.3 11494	*Ten.* Priore UNH, *ten.* Seeni WIA, *Serg.* Magnino UNH, *sold.* Mattioli UNH	Ground fire
11.8.1918	Feltre	Ca 11494	Lt. Park UNH, *serg.* Magnino UNH, *ten.* Contarini KIA, *sold.* Mattioli	Ground fire

Left: Without armament, Ca.3 2380 was used for liaison flights to occupied Austria in 1919. On the nacelle Vienna, Prague, and Pressbug (Bratislava) have been painted. The officer is *ten. oss.* Scotti Fantoni. (Archive Fiorenzo Longhi)

Below: Ca.3 4046 after the war went to Rome and was used for propaganda flights. Here it is seen with a Chinese officer. This airplane was then converted back into a bomber and served until 1925.

14ª Squadriglia

Above: The insignia of the 14ª Squadriglia.

Above: The pilot *tenente* Alessandro Petigax, who died in action on 26 September 1917, wore glasses.

The 14ª Squadriglia was formed at the Centro Formazione Squadriglie of Ghedi in November 1916 with *cap.* Giacomo Barucchi as commander. On 24 December one of its Capronis side-slipped after take-off and crashed injuring the pilots *cpr.* Scordamiglio and *cpr.* Giovanni Magnino. On 25 January 1917 the unit moved to Aviano, assigned to the IV Gruppo. Its personnel consisted of the pilots *cap.* Barucchi, *ten.* Beniamino Messeni, *s.ten.* Amedeo Digerini Nuti, *s.ten.* Carlo Toselli, *m.llo* Lorenzo Pica, *sergenti* Luigi Panizzon, Mario Frigerio, Tito Ragnelli, *cpr.* Artorige Galvani and *sold.* Antonio Cominotti; with the gunners Alfredo Cipriano, Andrea Leggero and Andrea Ruzzi. On 2 March the unit moved to Campoformido, equipped with Caproni 1190, 1238, 2606 and 2612.

The 14ª Squadriglia was engaged in the actions in preparation to the 10th Offensive of the Isonzo, flying its first combat mission on 13 April 1917, bombing Vippacco. During another mission on 21 April against the station of Pravcina, the crew of Caproni 2606, *s.ten.* Toselli and *m.llo* Pica, gunner Guglielmo Zanchi, noticed that one of the bombs was stuck in the wires of the main landing gear. Toselli tried in vain to get rid of the bomb, the pilots decided to land in the lagoon of Venice, where they were not recognized, so artillery started firing at them. A boat picked up the airmen and brought them to Caorle. The Command of the Group commended their behavior.

In May 1917 the pilots converted to the new Caproni 450 hp and trained in night flying using Ca 1211, assigned for that purpose to their airfield. This training was used on 30 June, when *cap.* Barucchi on his own initiative decided to bomb the arsenal and port of Trieste as a reprisal for an Austrian attack against Udine.

In the summer its personnel consisted of pilots *cap.* Barucchi, *ten.* Messeni, *s.ten.* Luigi Aquilanti, Digerini Nuti, Toselli, *m.llo* Pica, *sergenti* Frigerio, Galvani, Panizzon, Ragnelli, *sold.* Cominotti, observers *tenenti* Alberto Comellini and Ugo Degani, gunners *soldati* Antonio Cassoni, Cipriani, Osvaldo Gradara, Leggero, Pietro Pezzoni, Ruzzi, Guglielmo Zanchi

In August a *sottogruppo* (sub-group) was formed with an autonomous command at Campoformido and the 14ª Squadriglia was directly assigned to it, with the 15ª Squadriglia. In this period the unit used Caproni 450 hp 2316, 2318, 2319 and 2405.

In the late summer of 1917, the 14ª Squadriglia supported the 11th Offensive of the Isonzo. In one of such missions, a bombing over the Hermada Hill, on 5 September, Ca 4077 with pilots *ten.* Maurizio Petigax and Aquilanti, gunners Leggero and Maturino Sanzari, had its right engine radiator

Above: The Caproni of the 14ª Squadriglia, probably at Campoformido.

pierced by shell splinters. Leggero climbed over the wing and fixed the damage, and the bomber returned to Campoformido with all its engines running. On 20 September a complete crew was assigned to form the A.R. Detachment, thus *ten.* Taliedo Digerini Nuti, *serg.* Galvani, the observer Degani and the gunner Pietro Pezzoni left for Taliedo.

The 14ª Squadriglia suffered its first casualties on 26 September, when the Caproni 4066 of *ten.* Petigax, *ten.* Aquilanti, *cpr.* Leggero and *sold.* Sanzari failed to return from a mission to Grahovo. A shell destroyed the Caproni before it could drop its bombs, it fell down in flames with no survivors. While Italian reports credit this loss to artillery, however, there is a claim for two Capronis shot down by *Leut.* August Raben of Jasta 39, that had arrived on the Italian front nine days before.

On 4 October 1917 the 14ª Squadriglia stopped belonging to the *sottogruppo* and was re-assigned to the IV Gruppo. On 10 October, five days after the raid to Cattaro, Ca 4149 of Digerini Nuti, Galvani and the observer *ten.* Giovambattista Lucci Chiarissi, of the 9ª Squadriglia, fell in the sea near Monopoli, in Apulia. Galvani was killed, and as that was a flight without permission, *t.col.* Armani, the commander of the Distaccamento A.R. was officially reprimanded.

During the Austro-German offensive the 14ª Squadriglia was heavily engaged against the attackers, bombing the enemy at Grahovo. On 26 October the unit was ordered to retreat to Verona, but weather precluded flights, so with much sorrow the personnel torched Capronis 4077, 4143, and 4138, that were perfectly serviceable; only 4069 and 4139 were saved. On 4 November the unit moved to Ghedi, and then on to Padua, where weather conditions precluded any action, except one single sortie on 16 December, against the station of Grahovo.

On 24 December 1917, following the decision to shift the 14ª Squadriglia on the French Front, the unit was assigned to the newly-formed XVIII Gruppo and moved to Taliedo to be re-equipped. During the whole of 1917 the 14ª Squadriglia had flown 140 combat sorties.

The squadron left Taliedo by railway on 9 January 1918 and arrived in France two days later, settling on its new base of Longvic airfield, near Dijon. Still with *cap.* Barucchi as commander, it had the pilots *ten.* Digerini Nuti, *sergenti* Cominotti, Frigerio, Pica, Ragnelli, Mario Cazzaniga, Carmelo Cavalieri and *cpr.* Secondo Micheli; the observers *tenenti* Lucci Chiarissi, Claudio Lamberti, Italo Migliasso and Arturo Roseti; the gunners Gradara, Pezzoni, Oscar Roncati, Edgardo Scavizzi, Cesare Trocelli and Guido Vaccari. Their airplanes, Caproni 4086, 4232, 4234 and 4236 followed them on crates.

As soon as they were assembled, test flights began on 22 January. On 15 February 1918 the unit flew to Ochey airfield, its new base, and from there it launched its first combat mission on the new theatre on 17 March.

In March and April weather curtailed the number of missions, but the 14ª Squadriglia, operating always at night, bombed the railroad station at Thionville and the airfield at Montcornet, earning commendations from the Italian and French commands. In April it moved to a new airfield, Villeneuve, near Chalons sur Marne. In the night of 20 May, Caproni 4232 crashed on landing while returning from a bombing of the station of Montcornet, to which other Capronis of the XVIII Gruppo took part. The observer *ten.* Migliasso was killed, the pilots *sergenti* Cavalieri and Cominotti were badly injured, the latter dying a short time later.

At the end of May 1918, the 14ª Squadriglia was ordered back to Italy. On 10 June the unit was back at the Centro

Above: Caproni 2612 of the 14ª Squadriglia under line maintenance.

Formazione Squadriglie of Ghedi , there to be re-equipped with the 600 hp Caproni 5, its force growing from four to eight airplanes and from eight to 16 pilots.

On 23 July *cap.* Vito Pugliese, coming from the 15ª Squadriglia, became the new commander. Some documents report that some American pilots were assigned to this unit: 1st Lieuts. Thomas L. Orr, Clifford S. Andrews, Herbert L. Aman, Curtis A. Keen, Paul M. Potts Jr., and 2nd Lieuts. Oscar B. Marshall and Earl D. Forsythe.

The unit received eight new Caproni Ca.5 600 hp, losing one, Ca 11624 on 1 September 1918 during a ferry flight, and on 20 October 14ª Squadriglia returned to France, flying its Ca.5 over the Alps, back at Longvic, its first base in the previous assignment. Five days later it flew to Chermisey, and remained there until the end of the war, without flying any combat mission. The 14ª Squadriglia returned to Italy in early 1919, having however a fatal accident near Chermisey on 11 March, causing the death of *asp.* Pica, *serg.* Micheli and *ten.* Lucci Chiarissi. Lucci Chiarissi's brother Umberto, an artillery officer, had also died in France. The unit was disbanded on 17 November 1919. During the war it had flown 168 combat sorties.

Casualties and Combat Losses of the 14ª Squadriglia

Date	Place	Aircraft	Crew	Cause
26.9.1917	Grahovo	Ca 4066	*Ten.* Aquilanti KIA, *ten.* Petigax KIA, *cpr.* Leggero KIA, *sold.* Sanzari KIA	Ground fire? Fighters?
20.5.1918	Ochey	Ca 4232	*Serg.* Cominotti KIC, *serg.* Cavalieri WIC, *ten.* Migliasso KIC	Accident
11.3.1919	France	Ca.5	*Ten.* Pica KIC, *serg.* Micheli KIC, *ten.* Lucci Chiarissi KIC	Accident

Left: Ca.1 1211 was assigned to the 14ª Squadriglia for training in night flying.

Above: Ca.3 2319, even with all crew on board, is still sitting on its back skid.

Above: Ca.3 2612 of the 14ª Squadriglia.

Above: Ca.3 4609 of the 14ª Squadriglia armed with two 260-mm bombs, weighing 240 pounds each, and with mud guards at the wheels. (Archive Caliaro)

Above: Ca.3 4236 of the 14ª Squadriglia, in blue-grey night finish, with three American officers. It is not known if this photo was taken in France or in Italy.

Left: Ca.5 11624 was assigned to the 14ª Squadriglia, but did not fly on operations. Archive Longhi – Mario Chirico)

Above: Ca.5 11553 of the 14ª Squadriglia.

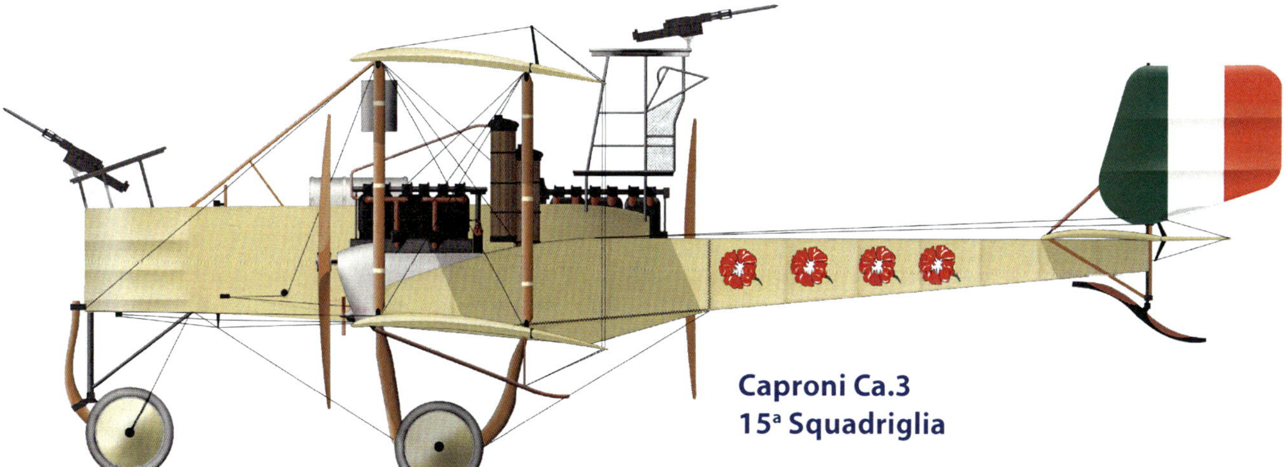

Caproni Ca.3
15ª Squadriglia

15ª Squadriglia

Above: The insignia of the 15ª Squadriglia

Formation of the 15ª Squadriglia began in December 1916 at the Centro Formazione Squadriglie of Ghedi. On 10 April 1917 this unit was at Aviano, ready for operations, assigned to the new XI Gruppo, with *cap. pilota* Leonardo Nardi commander. It was equipped with Caproni 450 hp serials 2342, 2343, 2345 and 2350. Its force included the pilots *tenenti* Arrigo Abba, Renzo Crescenzi, *s.tenenti* Domenico De Lorenzi, and Raul Silvagni and *serg.m.* Stefano Righi; the gunners Luigi Carletti, Alfredo Fantoni and Aldo Resca. First combat mission was flown on 17 April, a bombing of Chiapovano, and five more missions were flown in the month.

As the 10th Italian offensive developed, from 12 May to early June 1917, the 15ª Squadriglia was in action against targets at Vojscica, near Aisovizza and at S. Lucia. Its Capronis often returned with damages from shell splinters or from fighters' bullets, but the unit suffered no casualties. Also in the summer, during the 11th Battle, the squadron operated in the Carso region, bombing the stations of S. Daniele and Assling, Comen, barracks at Idria and on Hermada Hill. Its airplanes also took part in the mass night bombing raids against Pola in early August.

On 26 September 1917 the 15ª Squadriglia moved to Campoformido and there it formed the local Sottogruppo Squadriglie da Bombardamento with the 14ª Squadriglia. From there it sent a selected crew to Taliedo to form the Distaccamento A.R., consisting of *cap.* Nardi, *s.ten.* Silvagni, *sold.* Fantoni and the observer *ten.* Giorgio Morigi. The *Sottogruppo* (sub-group) was short-lived, ending on 4 October, the 15ª Squadriglia remained at Campoformido, now assigned to the IV Gruppo.

The collapse of Caporetto forced this unit to retreat to different airfields, it left Campoformido with 2342, 2343, and 4142 while Caproni 2345 was torched, as there was no pilot to ferry it. On 27 October it was at the Piazza d'Armi (parade ground) of Verona, then it went to Ghedi in the XIV Gruppo, finally on 17 November it was assigned back again to the IV Gruppo, which meanwhile had moved to San Pelagio. During the retreat it had operated against the advancing enemy. On 17 November it was tasked to bomb Primolano, taking off from Ghedi and then landing at San Pelagio, its assigned new base. Caproni 2342 with pilots *s.ten.* Vincenzo Biggio and *serg.* Riccardo Borello, observer *ten.* Federigo Nencini and gunner Rosario Ragonesi, suffered a failed engine and turned back with its bombs, landed too far up the landing field and crashed in a ditch at the end of San Pelagio runway. The bombs did not explode and the crew got out unharmed from the damaged bomber.

On 3 December Caproni 2343 of *cap.* Nardi and *sold.* Luigi Vallerga, observer *ten.* Dolores Lora Totino and gunner Gennaro Cuda, had a weird accident, when a magazine of its machine gun slipped away and broke the central propeller, its splinters damaging the flight controls. The pilots managed an emergency landing at S. Pietro in Gu, where the Caproni was later repaired and recovered.

On 24 December 1917 the 15ª Squadriglia was assigned to the new XVIII Gruppo, destined to France, and three days later it moved to Taliedo. During 1917 the 15ª Squadriglia had flown a total of 124 combat sorties.

On 11 January 1918 the personnel of the Squadriglia received Caproni serials 2366, 4053, 4225 and 4235 that had been shipped by rail to Longvic airfield, near Dijon. Under the command of *cap.* Nardi there were the pilots *tenenti* Abba, Biggio, Federico Molteni, *s.ten.* Righi, *sergenti* Raimondo Battistella, Borello, Vallerga and *cpr.* Vito Pozzi; the observers *tenenti* Lora Totino, Nencini, Giorgio Morigi and *s.ten.* Antonio Passagni; the gunners Cuda, Fantoni, Ragonesi, Luigi Carletti and Ezio Grassi.

The unit flew its first flight in France on 9 February, when an airplane dropped flowers over the City Hall of Lyon, in the course of a patriotic ceremony. On 19 February *cap.* Nardi returned to Italy and *cap. pilota* Vito Pugliese replaced him as commander.

On 28 February 1918 the 15ª Squadriglia flew to its new base, Ochey airfield, and from there on 17 March it sent out its first combat mission in the new theatre of operations. It was a moonless night sortie and one of the two bombers, Ca 4235, bombed the environs of Metz, as tasked, but on

Above: Ca.3 2343 of the 15ª Squadriglia having suffered a small mishap.

the way back it got lost in bad weather. The pilots, failing to find Ochey, force landed in a plain near Mussy sur Marne. In the rough landing the Caproni broke its landing gear and turned over; *ten.* Abba was slightly injured but the other crew members were unhurt. On 27 March two airplanes dropped eighteen 162-mm bombs over the German airfield at Frascaty, near Metz.

On 2 April 1918 the 15ª Squadriglia moved to Villeneuve. From its new base it attacked German targets at Laon, Malmaison, Montcornet and the station of S. Quentin. Molteni and Righi flew up to two sorties in a single night. In early June the Squadriglia with commander *cap.* Pugliese had Capronis 2336, 4165 and 4225. Its pilots were *tenenti* Abba, Biggio, Molteni, Righi, *sergenti* Battistella, Borello, Pozzi, and Vallerga; its observers were *tenenti* Lora Totino, Morigi, Nencini and Passagni; its gunners: Cuda, Grassi, Ragonesi, Vincenzo Cannoni and Luigi Carletti. On 12 June *cap.* Pugliese left, *ten.* Morigi became acting commander until 30 August, when *cap. pilota* Ferdinando Rospigliosi took over command.

On 24 July German airplanes bombed and strafed the airfield, killing some men of the ground services. *Ten.* Biggio and *ten.* Righi were undaunted, however, and they took off for their mission in the middle of the enemy attack. In the night of 29 July again *ten.* Molteni with *serg.* Vallerga took off while an attack was taking place: a German aircraft attacked the Caproni, that was hit but maneuvered to duel with its assailant, and then flew on to complete its mission.

On 7 August, the unit had moved to Chermisey, in the Department of Vosges, because the German had identified the airfield of Villeneuve and they were bombing it heavily. In the summer the unit bombed Laon, Fismes, Thionville and Linguyon. Il 14 October the 15ª Squadriglia, leaving some of its men on the French front, returned to Italy to convert on the Caproni 600 hp, at the Centro Formazione Squadriglie of Ghedi, where it was when the armistice was signed. The 15ª Squadriglia performed 233 combat flights during the war.

On 4 February 1919, with 16 pilots, 8 observeres and 8 aircraft, it moved to Poggio Renatico, and then in June 1919 it was assigned to the Comando d'Aviazione della Dalmazia, based first at Zara (Zadar) and then at Zemonico (Zemunik Donji) showing the flag in the region that was assigned to Italy despite the objections of the new Kingdom of the Yugo-Slavs. It was finally disbanded on 27 September 1919.

Above: The same airplane, from another angle, showing that it carried the red flowers, insignia of the 15ª Squadriglia, on the fuselages and a pennant on the central nacelle.

Above: Four flowers mark the fuselages of a Caproni of the 15ª Squadriglia at Aviano.

Above: A Caproni of the 15ª Squadriglia in Italy. (Archives ECPAD)

Below: Ca.3 2366 was an old airplane, that was refurbished and sent to France, in dark night finish. (Archive Roberto Bassi)

Ca.3 4225 of the 15ª Squadriglia in France; *ten. oss.* Lora Totino shows a 162-mm and a 260-mm bomb.

16ª Squadriglia

Above: Night take-off of a Caproni of the 16ª Squadriglia at Arquà Petrarca.

The 16ª Squadriglia was formed at Padua airfield on 22 January 1918, with commander *ten.* Nino Bartoli. It had three flights, the first one with S.P.3s armed with Fiat 25-mm cannon, and the other two with Capronis, respectively with 25-mm and 37-mm cannons. The flight using the SP.3, with pilots *ten.* Bartoli and *sergenti* Angelo Coralli, Leonardo Cristofani and Pietro Dao, was tasked with protecting Padua from airships and it reported to the Comando Difesa Aerea of that city, while the Caproni flights had offensive tasks against targets such as trains, stations, carriages and troops in march, and it reported to the Comando d'Aeronautica of the Supreme Command. The Caproni pilots were *ten.* Tommaso Salazar, commander, Antonio Crudeli, *ten.* Eugenio Pasquino, *serg.* Giovanni Redivo *cpr.* Luigi Po and *sold.* Gildo De Santis,

On 24 March 1918 during a ferry flight from Taliedo, Caproni 2406 was lost, pilots *s.ten.* Antonio Crudeli and *serg.* Giacomo Rossi were slightly injured.

The squadron flew its first combat sortie on 29/30 March, attacking Primolano with S.P.3 4550 and 4654 of Bartoli and *ten.* Tommaso Salazar, Dao and gunner Falcioni. Both airplanes crashed in their return, but the airmen were unhurt.

On 12 April 1918 the unit moved to Arquà Petrarca airfield and two days later it was assigned to the XIV Gruppo. The Capronis flew their first war mission on 14 May, bombing Feltre airfield, probably having given up the use of the cannon. Ca 2404 with pilots Salazar and *serg.* Luigi Pò, observer *ten.* Balbiati and gunner Carlo Surniotti dropped 12 bombs.

On 24 May the S.P. 3 flight with aircraft 4628 and 4536 left, the 16ª Squadriglia then remained with Capronis 2314, 2401 and 2404. Its commander was Salazar, with pilots Crudeli, Pò, Rossi, *ten.* Eugenio Pasquino, *serg.* Giovanni Redivo and *sold.* Gildo De Santis; gunners Surniotti, Carlo Agnello, Antonio Crifò, Secondo Fassio, Mario Mugnaini, Luigi Pola, Italo Soprani and Giuseppe Zamolo. On 1 June it bombed the Austrian airfield at S. Pietro in Campo and on 7 June the airfield at Godega di Sant'Urbano.

The Comando Supremo disbanded the 16ª Squadriglia on 20 June 1918, passing its equipment and personnel to other units. The 16ª flew a total of six combat sorties, and one *Medaglia di Bronzo* was assigned to Surniotti.

Above: Cannon-armed Caproni of the 16ª Squadriglia.

Above: SP.3 4628 of the 16ª Squadriglia at Arquà Petrarca.

Distaccamento A.R.

Above: A Short Type 320 arned with a torpedo, hanging from a crane at Brindisi. The raid on Cattaro on 2 September 1917 failed due to bad weather.

Cattaro (today Kotor, Montenegro) was one of the most important and best protected Austrian naval bases, a deep bay connected to the sea by a narrow channel. No naval attack could violate that secure base, so the Allies started considering aircraft. The Royal Navy had been the first to entertain such a project in August 1917. They had a complex plan, with six Short Admiralty Type 320 floatplanes armed with torpedoes to be carried on rafts close to the target, sailing from Brindisi, and the action took place on 2 September 1917, but a storm made it impossible to launch any of the aircraft and the raid was cancelled.

The Italian Supreme Command considered the use of Caproni 450 hp bombers. Gen. Maggiorotti, the Chief of Aeronautical Services, in a document dated 28 August 1917 considered using 16 bombers, modified for a flight of six hours, taking off from Gioia del Colle, Apulia, benefiting from the moonlit period from 25 September to 8 October. The Regia Marina was to have the responsibility of the whole operation, determining the route and providing support at sea, while the Army was to provide searchlights for the airfields of Gioia and Foggia.

Difficulties and risks of this operation were obvious: a flight of 500 km, 400 of them over sea, with a landplane that had no possibility of ditching in the water. Even if a Caproni had alighted correctly it could float for no more than one minute, leaving its airmen to wait in the sea at night some sort of rescue from a ship. Navigation itself was a problem, with the primitive instruments then available.

A special Group was formed for this operation, consisting of two Squadriglie, 1ª bis and 15ª bis, under the command of *magg. pilota* Armando Armani, and volunteers were so many that they had to draw lots to participate. The first one obviously was Gabriele d'Annunzio, who had been pressuring in order to be accepted. The Gruppo was called Distaccamento A.R., from the initial letters of the name of its commander; d'Annunzio was commander of the 1ª bis, *cap.* Leonardo Nardi of the other Squadriglia.

The crews gathered at La Comina and from there they went to Taliedo, Milan, to get directly from the Caproni factory the new bombers specifically prepared for this operation. They received Caproni Ca 450 hp serials 4146, 4148, 4149,4150, 4151, 4152, 4154, 4155, 4157, 4158, 4159,

Above: Ca.3 4158 was produced for the bombing of Cattaro.

4160, 4161, and 4162.

Under the command of Armani there were the best men among the Italian bombers, the pilots *capitani* Leonardo Nardi and Maurizio Pagliano, *tenenti* Mario Giulio Agostini, Casimiro Buttini, Giampiero Clerici, Amedeo Digerini Nuti, Luigi Gori, Mario Muratorio, Mario Panciera di Zoppola, Raffaele Paravicini, Giovambattista Pittaluga, Vito Pugliese, Edoardo Scavini, Raul Silvagni, *sottotenenti* Igino Bevilacqua, Federico Molteni, Gino Lisa, Ivo Oliveti, *aspirante* Cesare Baccili, *sergenti* Antonio Angelelli, Artorige Galvani, Armando Manicardi, Angelo Pacassoni, Battista Origgi, Marco Resnati, Arturo Vischioni and *cpr.* Riccardo Casellato; the observers *cap.* Gabriele d'Annunzio, *tenenti* Paolo Benigni, Domenico Corica, Mariano D'Ayala Godoy, Fausto Da Passano, Ugo Degani, Giovambattista Lucci Chiarissi, Giorgio Morigi, Weiss Poccetti, Alberto Scalcerle, Guido Toffoletto, Raffaele Vallarino and Attilio Zoppi; the gunners Mario Catapan, Alberto Farneti, Beniamino Lattanzio, Agostino Montaldo, Pietro Pezzoni, Enrico Pucci, Paolo Raccagni, Primo Rapetti, Emilio Saletta, Battista Spreafico and Enrico Vanzulli. And there was also one civilian, Guelfo Civinini, war correspondent of the newspaper *Corriere della Sera*. Civinini was a civilian, yet during the war he earned four *Medaglie di Bronzo al Valor Militare*.

On 25 September 1917 these 14 airplanes flew to Centocelle, Rome. They left the capital the next day and they all reached Gioia del Colle without any problem. They were carefully prepared for the mission, scheduled for the night of 29 September. They had liquid compasses similar to the ones of the torpedo boats of the Navy, which *Tenente di Vascello* Andrea Bafile carefully adjusted. Bafile, commander of the torpedo boat "*Ardea*", temporarily relinquished his command and contributed to the preparation of the raid, flying as observer on Caproni 4148 of Baccili, Pacassoni, and Raccagni. Five months later, leading a group of sailors deployed on the lower Piave, he was killed during a sortie near the Austrian lines, trying to recover a comrade. He was posthumously assigned the *Medaglia d'Oro al Valor Militare*.

Airplanes and men were ready, but there were no bombs, which were coming by railroad and were lost somewhere in transit along the clogged southern lines. The missed ordnance arrived in the morning of the 30th, but meanwhile weather had deteriorated, with clouds and rain. The morale of the crews was suffering, as the window of opportunity was narrowing. Finally on 3 October the weather forecast was good, and yet the prudent Armani decided to wait, raising the ire of the hot-tempered d'Annunzio.

On 4 October the day was good, the go-ahead was given for the evening and the final checks were performed. At 11 p.m. the first airplane, Ca 4158 of Armani, took off, followed four minutes later by 4160 of Pagliano and Gori, with the observer d'Annunzio. Civinini was on Ca 4151 of Nardi,

Although the bombing of Cattaro received wide coverage, no photographs seem to have been taken. This is the only known image of the event, showing d'Annunzio in the cockpit of Ca 4160.

Above: Italian personnel, among them a British officer, crowding around Ca 4162.

Silvagni and Morigi. One by one the bombers took off and followed the route marked by searchlights. At sea a similar line of searchlights was expected from torpedo boats placed at intervals of 30 km, but on account of the haze the crews could spot only the first one, and then they progressed only navigating with the compass and the stars.

After one hour Caproni 4162 of Muratorio, Angelelli, Benigni and Spreafico was forced back because of engine troubles, and then Ca 4146 of Buttini, Casellato, Poccetti and Farneti also had to interrupt the mission and turn back. The other ones went on in the darkness and broke land slightly south of their target, as the wind had caused them to drift. They found their position and flew individually. From an altitude of 3,000 meters, they dropped a total of 24 260-mm mines and 72 162-mm bombs over the mooring and the docks, setting fire to some warehouses at Cienvich and to a fuel tank at Kumbor. Anti-aircraft reaction was light and badly coordinated, proving that their operation was a total surprise.

The return flight followed just the compass and the stars, because the torpedo boats were still invisible. All the Capronis but one returned safely to Gioia del Colle before daybreak. The

Above: After their mission, some Caproni visited the Foggia flying school, for the morale of the American trainees.

missing one, Ca 4154 of Paravicini, Resnati and Toffoletto, landed without damage in a field near Manfredonia.

This operation was mentioned in the Bulletin of the Supreme Command. The Chief of Staff of the Navy gave to all the officers and the sergeants who had participated a *Medaglia di Bronzo,* but only an official praise for the soldier gunners.

Before the Distaccamento was disbanded and its personnel returned to their Squadriglie, there were casualties. On 10 October *serg.* Galvani of the 14ª Squadriglia was killed in an accident. Ca 4149 that he was piloting with *ten.* Digerini and carrying *ten.* Lucci Chiarissi got lost and fell in the sea off Monopoli after a flight over Taranto without permission. Armani was officially reprimanded for having allowed a flight *"with no military necessity and in an hour that was unfit on account of weather conditions".*

On 15 October there was another accident to Ca 4154, that was flying to test at sea some new type of bombs. Just after take-off, the right engine quit and the pilots managed an emergency landing against a strong crosswind. The airplane drifted considerably, smashed its landing gear and rolled over. *S.ten.* Antonio Lucciola, of the 255° Battaglione Milizia Territoriale, was killed, while *tenenti* Pugliese and Paravicini suffered only minor bruises. Lucciola was on board without permission, as Paravicini had just taken him for a joyride.

In the final ten days of October the crews returned to their units, just in time to go into action after the Caporetto breakthrough, many of them to be killed in that period.

General Maggiorotti had contemplated five night bombing missions in that moonlit period. Weather conditions, lack of bombs and other factors allowed only one. Still, although an isolated event, the operation of Cattaro had a strong repercussion and it was an important testimonial of the capacity of the Italian bomber force to attain strategic objectives.

Losses of the Distaccamento A.R.

Date	Place	Aircraft	Crew	Cause
10.10.1917	Monopoli	Ca 4149	*Ten.* Digerini UNH, *serg.* Galvani KIC, *ten.* Lucci Chiarissi UNH	Accident
15.10.1917	Gioia del Colle	Ca 4154	*Ten.* Paravicini UNH, *ten.* Pugliese UNH, *s.ten.* Lucciola KIC	Accident

Above: American pupils crowding around Ca 4155 at Foggia. (Archive Apostolo)

Above: The 36th Fighter Wing of the Italian Air Force, based at Gioia del Colle, erected this monument in 1969.

Made in United States
North Haven, CT
04 November 2026

21731730R00146